SUBSTANTIAL SUBSTRATA

VS.

INSUBSTANTIAL SUBSTRATA

✝✡✝✡✝✡✝✡✝✡✝✡✝✡✝✡✝✡✝✡✝✡✝✡✝✡✝✡✝✡

SUBSTANTIAL SUBSTRATA

vs.

INSUBSTANTIAL SUBSTRATA

BY

EDWARD N. HAAS

ISBN: 1-4033-3958-9 (e-book)
ISBN: 1-4033-3959-7 (Paperback)

Library of Congress Control Number: 2002092372

This book is printed on acid free paper.

Printed in the United States of America
Bloomington, IN

1stBooks – rev. 08/01/02

ABOUT THE BOOK

Imagine holding, in the palm of your right hand, a smooth, solid glass sphere about 4 inches in diameter. It's mostly transparent but tinted blue. In the palm of your left hand, is another such globe tinted red. From of old, many have often said that, whatever else they might be, each of those spheres is a bundle of the substratum called matter, and each of those bundles is a separate thing because it is to no extent stuck either to the other globe or to your hand or to any other bundle of matter. They then went on to add 2 other crucial characteristics.

In the first place, they said that, being a substratum, each of these balls is that to which we can attach predicates, but which we cannot sensibly use as a predicate. For example, of the sphere in your right hand, we can sensibly say: "This globe is blue", but cannot sensibly say: "This blue is globe." Again, we can sensibly say: "This bundle of matter is glass", but cannot sensibly say: "This glass is bundle of matter." We most certainly cannot say: "This blue globe is this red globe."

In the second place, they said that this substratum called matter is—as are all substrata—able to avoid annihilation even without being stuck to some other substratum. For example, the blue globe does not have to cohere to the red globe or to your hand in order for its *matter* to avoid disappearing into thin air without a trace. Oppositely, the instant you shatter the blue globe, its spherical *shape* will utterly vanish.

When one says of some substratum that it _is_ able to avoid annihilation even without being stuck to some other substratum, one effectively calls it *substantial*. If one says of some substratum that it is _not_ able to avoid annihilation without being stuck to some other substratum, one calls it _in_substantial.

Say all substrata are *substantial*, and, when it's time to explain the relationship between the glass globes you see and the ultimate sub-atomic particles racing around in each globe's atoms, you'll describe it one way. You'll say those ultimate particles <u>are</u> the globe's matter, and that makes it impossible to explain whence come either globe's *sensible* characteristics. For, there is no comparison whatsoever between those *sensible* characteristics and the characteristics of what you have branded the only pieces of matter ever to be found anywhere. But, if the sensible characteristics are not the internal characteristics of *the only matter you acknowledge*, then of what **are** they the internal characteristics? Nothing?! Say some substrata are *in*substantial, and you'll describe that relationship very differently. You'll say each globe's matter is something cohering to, and ever being renewed by, those ultimate particles and, thereby, filling the gaps between them. That makes it easy to explain the globe's sensible characteristics. For, you've then introduced a new kind of matter which may well have internal characteristics identical to those you sense.

Is there such a thing as an insubstantial substratum, or are all substrata substantial? What are the consequences of saying that all substrata are substantial?

This book answers those questions in what is perhaps obnoxiously laborious detail. In the process of doing so, it seeks to achieve two goals: (1) To show how universal skepticism is the necessary result of saying all substrata are substantial, and (2) to lay the foundations for the most novel cosmological theory in history—a theory which some have called "a thing of beauty like no other I've ever witnessed."

In recognition of the fact that
The happiest days of my life were spent at
The Discalced Carmelite community at
Oakville, California,
I gratefully dedicate this work to:

The Fathers And Brothers
Of
The Discalced Carmelites
Of
The Province Of California.

BOOKS IN PRINT BY EDWARD N. HAAS

The Lovesong Tree, A Fairy Tale Portrait Of God

The Story Of Drawden The Pig

Introspective Cosmology II

Letters And Thoughts On Homosexuality

The Nature And Origins Of Murder Worship, The Ultimate Disease

Pieces Of Moral And Dogmatic Theology

In The Beginning Was The Internet

Two Letters For 1993

Letters Against Murder Worship

A Letter From A Father To His Son In 1994

On Philosophy: One Long And Four Short

Miscellaneous Letters

Substantial Substrata Vs. Insubstantial Substrata

CONTENTS

DESCRIPTION   PAGE NO.

i

CONTENTS
(Continued)

DESCRIPTION   PAGE NO.

CONTENTS
(Continued)

DESCRIPTION   PAGE NO.

CONTENTS

(Continued)

DESCRIPTION   PAGE NO.

CONTENTS
(Continued)

CONTENTS

(Continued)

DESCRIPTION   PAGE NO.

CONTENTS
(Continued)

DESCRIPTION   PAGE NO.

LATE FOOTNOTES

CONTENTS
(Continued)

DESCRIPTION   PAGE NO.

LATE FOOTNOTES
(Continued)

CONTENTS
(Continued)

DESCRIPTION   PAGE NO.

LATE FOOTNOTES
(Continued)

CONTENTS
(Continued)

DESCRIPTION   PAGE NO.

LATE FOOTNOTES
(Continued)

LEGAL NOTICE OF PERMISSION TO QUOTE IN A SCHOLARLY DISSERTATION

To every party of one or more persons composing—in voice, print, or electronic media and whether for profit or not—a single scholarly dissertation either upon some scholarly dissertation of mine or upon ideas relevant to ideas expressed by me in some scholarly dissertation of mine whether published or unpublished, I, Edward N. Haas, hereby grant the right to quote—without fee or further permission whether in writing or otherwise and regardless of the manner or the extent to which the quoted words might be crucial to their source—up to a total of 2,000 of my **_own_** words or translations as found in some scholarly dissertation of mine and to quote them anywhere in the course of the dissertation whether in its body, its footnotes, its endnotes, or as fill between chapters or as chapter headings. This grant also extends to charts, drawings, and figures created by me in the course of my scholarly dissertations and, where no word count is possible in such charts, drawings, and figures, each single chart, drawing, and figure shall, in that case, be counted as one hundred (100) words. To facilitate the word counting process, each complete number no matter how large or small shall be counted as one word, and each formula no matter how many or few its components shall be counted as one word.

This grant is valid provided only that the quoting party meets these following five (5) conditions: (1) What's quoted from me must be quoted in what is truly a scholarly dissertation and is one containing one or more ideas relevant to what is quoted from me (Lewd works posing as scholarly examinations into sexual behavior do not qualify.); (2) The source of my words, charts, etc. and the pertinent copyright notice are fully described; (3) in the grantee's scholarly dissertation, the ratio of: (a) the total number of words written by the grantee, to (b) the total number of quoted words written by myself, is no less than twenty to one (*i.e.:* 20/1); (4) in the act of availing himself (herself or themselves) of this grant, the grantee does thereby automatically and permanently extend—and forever bind all the grantee's heirs, assigns, publishers, and successors in title to grantee's copy rights to extend—to all other composers of scholarly dissertations the same privilege with regard to the *grantee's* scholarly dissertations as the grantee has accepted from me with regard to *my* scholarly dissertations; and (5) the grantee places near the front of the grantee's scholarly dissertation a statement (written if possible) acknowledging that the grantee accepts these conditions. As long as these five (5) conditions are met, this grant is permanent and irrevocable and remains forever binding upon myself, my heirs, assigns, publishers, and any successors in title to my copy rights.

Preface To The Edition Of 2002:

A. WHY THIS PREFACE:

To myself, the distinction between substantial and insubstantial substrata is as clear as a bell. Disgustingly, decades of trying to communicate that difference to others has proven to me, over and over again, that it is virtually impossible for anyone else to understand what I'm talking about.

As I begin this latest and perhaps futile effort to communicate that difference, a desperate thought occurs to me: Maybe at least some readers would find the going easier, if I were first to familiarize them with the most basic of the concepts comprising the cosmological theory which I have founded upon the distinction between substantial and insubstantial substrata. At 66, I'm so desperate to find a way to ease the burden of following my way of philosophizing, I might as well try it.

So then, the remainder of this preface shall present the reader with a synopsis of a cosmological theory which, in its original presentation in my book **Introspective Cosmology II**, took 570 pages and 238,000 words. Since what follows takes less than 7,000 words, it's quite an achievement in the art of synopsizing.

B. FOUNDATION

What is the difference between scientific and philosophical cosmology? The scientist turns to the world outside of his mind, uses a multitude of instruments to observe and measure perhaps millions of natural phenomena, and then tries to formulate a theory which gives order to what he's found in that extra-mental world. The philosopher turns to his armchair, reflects inward upon his own mind alone, and then tries to give order to what seem the most basic thoughts he's found in that intra-mental world.

This philosophical cosmology began over forty years ago when its author noted that there is a tri-partite structure necessary to every act of consciousness, namely: (1) that which is being aware, (2) that of which #1 is aware, and (3) the act which joins #1 and #2. That introspective observation quickly gave birth to the conclusion that every ultimate constituent (*i.e.:* every physically separated unit which does not, under magnification, prove to be merely a conglomerate in which even more basic constituents are physically separated from one another) is inherently triune (*i.e.:* is 3 subjects inseparably conjoined in one object). That quickly suggested a principle which the author originally expressed in the following quasi-algebraic fashion:

$$(+a) \otimes (-a) \rightarrow 0$$

Applying that first principle to itself resulted in a second principle which the author originally expressed so:

$$(+a) \otimes (-a) \rightarrow 0$$
$$\otimes$$
$$(-a) \otimes (+a) \rightarrow 0$$
$$\downarrow$$
$$0$$

Decades later, the quasi-algebraic expression of the first principle turned to this:

$$\uparrow$$
$$(+a) \otimes (-a)$$
$$\downarrow$$

In turn, that gave birth to this geometrical expression:

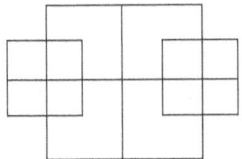

To understand what the geometrical expression is saying, one must first imagine that the squares are cubes. The large cube is *two* cubes rotating in opposite directions around their common center as each takes a small cube with it. Concurrently, the two small cubes, at a faster pace (and moving in and out from the large

3

cubes' center), are rotating around their own centers and doing so either in the same or opposite directions. In the latter case, the arrows in the quasi-algebraic expression would indicate neutrality on the axis of rotation. In the former, they would indicate either a positive or a negative "charge" on that axis, depending upon whether each of the small cubes is rotating clockwise around its own center or counter-clockwise.

That, then, is a rudimentary description of the foundation which gave birth to the concepts about to follow. Here again, what follows is very rudimentary, since 7,000 words are a mere shadow of 238,000.

C. FORMS:

The Haasian concept of a form can first be depicted as a larger, transparent cube with a smaller opaque cube centered inside the larger cube. As for *why* that is so, let's leave that till later. For now, be content to imagine that the two cubes are concentric.

The larger, transparent cube represents the *potentiality* of the form, and the smaller, opaque cube represents the *actuality* of the form. Imagine that the dimensions of the smaller cube are 50% those of the larger cube. That's a way of saying that, in its current act, the form is *actualizing* one-eighth of the potency available to it.

Next, imagine that what we've depicted thus far remains that way as we examine it for a while and call it: "stage one of a two stage cycle". After we have examined the picture to our heart's content, we snap our fingers, and, in an instant, the dimensions of the inner, opaque cube are suddenly the same as those of the larger, outer cube. In other words, we are now looking at a single, opaque cube. That's a way of saying that, in its current act, the form is *actualizing* all of the potency available to it. As we examine this new state of the form for a while, we call it: "stage two of a two stage cycle".

After examining it for a while, we snap our fingers, and, in an instant, the dimensions of the inner, opaque cube are once again one-half those of the outer, transparent cube, and we have returned to stage one of the two stage cycle. It's just that now we have commenced the second of those cycles.

So far, that's a way of saying that this particular form is capable of only two distinct acts, and, as we've seen, the difference between those 2 acts is this: In one act, the form actualizes 100% of the potency available to it, and, in the other, actualizes 12.5%.

In Haasian cosmology, that particular form is called: "the form of the universe", "the form whose native ontological distance is 2^{256} (*i.e.:* roughly 1.15792 x 10^{77})", and "Omega F1". Only one such form has been created *native* to that ontological distance (*i.e.:* OD for short), and "native OD" means the ontological distance at which a form was created. It also means something about how the form always rotates on two of its three axes. Well get to that later.

D. FORMS AND ONTOLOGICAL DISTANCE:

What, though, is "ontological distance"? For one thing, OD determines the description of every form's two basic acts while still at its native OD. From their *native* OD, forms can change to a *current* OD and, thereby, increase the number of different acts they can perform.

SUBSTRATA

To illustrate that, let's examine the form at the *opposite* end of the scale, namely: the form native to OD *one*. It's called "Alpha F1". Here again, there is an outer cube which is larger and transparent, and an inner cube which is opaque and much smaller. As before, their centers are one. As before, the outer cube of Alpha F1 is as large as the outer cube of Omega F1. For all forms, the dimensions of the larger, outer cube equal those of the form of the universe. That's a way to say all forms have the same *potentiality*.

What, now, shall we say of the dimensions of Alpha F1's smaller cube? As some may have guessed already, those dimensions are 2^{-256} (Notice the minus sign.) times those of its larger, outer cube.

Here again, as long as the form is at its native OD, it can perform only 2 distinct acts. In one, it actualizes $(2^{-256})^3$ of the potentiality available to it, and, in the other, it actualizes an eighth of that.

Should a form native to OD 1 move to a current OD of 2^{128}, it would then be capable of 2^{128} different acts. The dimensions of those acts would be 1, 2, 3, 4, 5 . . . 2^{128} times the dimensions of the smallest act.

In this 9th epoch of the universe (due to last 18 trillion Earth years), 6^{256} (*i.e.:* 1.61×10^{199}) forms have been created native to the ontological distance of 1, and, in Haasian cosmology, their primary function is to produce the electrons. The number of forms created native to OD 2 & 3 is 6^{255}; the number created at OD 4 thru 7 is 6^{254}; the number created at OD 8 thru 15 is 6^{253}, and so forth, until we have 6 created native to the distances of 2^{255} thru (2^{256})-1 and 1 created native to 2^{256}.

What shall we say of the forms created native to the ontological distances of 2, 3, 4, 5, etc.? What's the inner cube's size relative to the outer cube? If you've followed the above, it is easy to calculate it. Simply divide the OD of the form by 2^{256} to obtain the dimensions of the largest act, that by 2 to obtain the dimensions of the smallest act, OD^3 by $(2^{256})^3$ for the percent of potency actualized in the former, and OD^3 by $(2^{257})^3$ for the percent actualized in the latter.

E. 3 KINDS OF ONTOLOGICAL DISTANCE:

Ontological distances are of three kinds: categorical, generic, and specific. The 10 *categorical* OD's are: 2^0 (*i.e.:* 1), 2^1, 2^2, 2^4, 2^8, 2^{16}, 2^{32}, 2^{64}, 2^{128}, and 2^{256} (*i.e.:* successive squares of the powers of 2). The 246 *generic* OD's are: 2^3, 2^5, 2^6, 2^7, 2^8, 2^9, 2^{10}, 2^{11}, 2^{12}, 2^{13}, 2^{14}, 2^{15}, 2^{17}, etc.. In other words, every OD describable as a power of 2 is a *generic* OD, unless it is already a *categorical* one. All OD's which are neither generic nor categorical are *specific*. The significance of the 3 kinds will be explained later.

As long as all forms are still at their native OD, every form at and below the 9th categorical form (*i.e.:* the form at 2^{128} which is to say the square root of 2^{256}) is in what we would call the *micro*scopic world, and every form above 2^{128} is in what we would call the *macro*scopic world. That's because the OD of 2^{128} is what Haasian cosmology calls the "mirror threshold". All ontological distances lower than 2^{128} are on one side of the mirror, and those higher than 2^{128} are on the opposite side of the mirror. Naturally, what's on one side of the mirror is a reflection of what's on the other and, therefore, reverses it in many ways.

As long as there is only one form per OD, every form's center is the same, and the result is a largest, outermost box with a smaller box inside of it and a smaller box inside of it, and so forth. Forms can coincide that way because they are not *physically* outside of one another in *space*; they are *logically* and *actually* outside of one another in OD. That makes OD the 4th dimension of static space (*i.e.:* space without time as a factor).

F. 6 KINDS OF POTENCY:

If forms are merely *logically* outside of one another in OD, how is it possible to have several forms at the same OD? Before we answer that, let's explain why forms can be depicted as cubes. Every form is in potency to something which is *technically* (*i.e.: logically*) outside of the universe, and this something is a created, finite version of *The Infinite Form Itself*. The Infinite Form forever performs the same one act describable as: an act in which 100% of an infinite supply of potentiality is actualized.

The Infinite Form can be described as 1 "object" in which 3 "subjects" share the entire object with one another. These three subjects can be labeled A, B, and C, and each is equally related to the other two. But, A related to B is not *identical* to A related to C; B related to A is not *identical* to B related to C; and C related to A is not *identical* to C related to B. The result is that we have six distinct frames of reference: A, A' (*i.e.:* A prime), B, B', C, and C'. A = A related to B, and A and B are mutual opposites. A' = A related to C, and A' and C are mutual opposites. B = B related to A. B' = B related to C. C = C related to A, and C' = C related to B. B' and C' are mutual opposites. In other words:

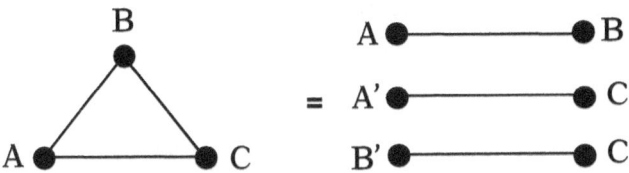

The *logical* way to depict all of that is to have three axes (*i.e.:* AB, A'C, and B'C') each of which is perpendicular to the other two and all equal in length. The logical result is a cube. Naturally, whatever is in potency to that cube is also a cube with *six* different kinds of potency, which we can label: "the A potency", "the A' potency", "the B potency", "the B' potency", "the C potency", and "the C' potency".

If you've followed that, we can now get a better picture of what a form's actuality means. For, what you should now be able to see is this: Whenever a form converts some of its potency into actuality, it's actualizing the same percentage of *six different potencies*.

To illustrate that, go back to the "form of the universe" (*i.e.:* Omega F1 a/k/a the form native to OD 2^{256}). Earlier, we said that, in one of its acts, it actualizes 12.5% of its potency and, in the other, actualizes 100% of its potency. We can now be more elaborate and say this: In one of its acts, it actualizes .125 of its A potency + .125 of it's A' potency + .125 of its B potency + .125 of its B' potency + .125 of its C potency + .125 of its C' potency and, in the other, actualizes 100% A + 100% B, etc.. In other words, in a given act, a given form actualizes each of its six potencies equally, and that's why the center of a form's *actuality* never moves from the center of its *potentiality*. As a result, though a form can move up and down in current ontological distance, no form can *logically* move the center of its actuality away from the center of a single, solitary one of the centers of all the other forms. How is such a limit to be overcome?

G. GENERATORS:

That brings us to what Haasian cosmology calls "generators". There are three primary differences between forms and generators. First: In any given one of its acts, a *form* actualizes each of its six potencies equally; in any given one of its acts, a *generator* actualizes, at most, *three* of its six potencies and does not necessarily do so equally.

Why is that? It's because no generator can—in the same one act—simultaneously actualize both a potency and its anti-potency. For example, if, in a given act, a generator is currently actualizing some percentage of its A potency, it cannot simultaneously actualize any of its B potency. Again, if, in a given act, a generator is currently actualizing some percentage of its A' potency, it cannot simultaneously actualize any of its C potency, etc..

To illustrate the effect of that limitation, imagine again a transparent cube representing the potentiality (*i.e.:* six potencies) of some given generator. In this case, though, the dimensions of this cube will not be the same as the dimensions of every *form's* outer cube. We shall explain later why that's the case. For now, the more important difference is this: Imagine that the outer cube is a stack of smaller cubes. For example, imagine a transparent cube composed of 4 layers each having 4 rows each containing 4 cubes. Each of those 64, smaller, transparent cubes now represents one of the 64 different acts which can be performed by the generator to which the cube applies.

How is that so? Well, one side of the cube represents the A potency while the opposite side represents the B potency (***i.e.:*** the anti-potency of A). A third side of the cube represents the A' potency while the opposite side represents the C potency. Yet a fifth side represents the B' potency while the opposite side represents the C' potency. If you follow all of that, then you should be able to see this: One of the 64 smaller cubes is in a corner where the A, the B' and the A' sides meet. If you can, picture it as the corner which is top left and to the rear (***i.e.:*** away from the viewer). If the center of the generator's *actuality* is at the center of that cube, then it is performing an act which can be described as follows: (100% A – 50% B) + (100% A' – 50% C) + (100% B' – 50% C'). But, aren't the percentages wrong? Be patient.

There are three cubes adjacent to that first cube, if we reject diagonals. One cube is to the right of that first cube; another is directly below it; and the third is in front of it toward the viewer. That is a *graphical* way of expressing which acts are *logically* adjacent to the act described by the first cube.

If, therefore, the generator's actuality is to change from the current act to another, it must switch from the first cube to one of the 3 adjacent to it. Why?! Remember: Forms change from act to act by merely changing the dimensions of the smaller, inner cube, which is to say the actuality of a form merely expands and contracts relative to its center at the center of the outer cube. Thus, for every form, the center of its actuality never moves relative to the center of its potentiality. The opposite is true of a generator's actuality: Without expanding and contracting relative to its actuality's center, every generator changes from act to act by shifting its center relative to the center of the outer cube. Thus, for every generator, the center of its actuality moves relative to the center of its potentiality with every change from act to act.

Let's assume the actuality of the generator moves from the first cube to the cube on its right. That's a way of saying it has ceased to perform the act described as (100% A – 50% B) + (100% A' – 50% C) + (100% B' – 50% C') and has now commenced to perform an act describable as (50% A – 0% B) + (100% A' – 50% C) + (100% B' – 50% C').

Now, let's explain the "erroneous" percentages. Imagine a cube which measures 1 inch on each side. Six strings each 2 inches long are protruding from the cube such that each string protrudes from the center of, and is perpendicular to, one of the cube's sides (Since I know nothing of "String Theory", I'm not trying to tell you that Haasian cosmology is a kind of "String Theory"). If you prefer, imagine 6 rigid pieces of wire. The strings (wires) are joined at the center of the cube; and so, that means each string is one-half inch into the cube, and its remaining 1.5 in. is sticking out of the cube. The string going to the left is the A string; the string going to the right is the B string (and this is the A string's anti-string); sticking out of the top of the cube is the B' string; sticking out of the bottom of the cube is the C' string (and this is the B' string's anti-string); the A' string is sticking out of the rear of the cube and away from the viewer; and the C string sticks out of the front of the cube and points at the viewer (and this is the A' string's anti-string).

If you've got that picture firmly in hand, shift the center of the cube one-half inch to the left along the A string. In that position, 50% of the A string (*i.e.:* 1 in.) is inside of the cube and 50% is still sticking out to the left, and, simultaneously, 100% of the B string (*i.e.:* 2 in.) is sticking out of the right side of the cube.

What of the other 4 strings?! They have moved along the A string with the cube's center. Thus, 25% (*i.e.:* .5 in.) of the A', C, B' and C' strings is inside the cube; 75% (*i.e.:* 1.5 in.) of the A' string is still sticking out the center of the rear; 75% of the C string is still sticking out the center of the front; 75% of the B' string is still sticking out the center of the top; and 75% of the C' string is still sticking out the center of the bottom.

If you've followed that, imagine the center of the cube shifts one-half inch away from you along the A' string. In that position, 50% of the A' string is inside of the cube; 50% is still sticking out to the rear; and 100% of the C string is sticking out of the cube's front and toward the viewer.

What of the other 4 strings?! They have moved along the A' string with the cube's center. That necessarily means that 50% of the A string is still inside the cube, 50% of the A string is still sticking out to the left of the cube, and 100% of the B string is still sticking out to the right of the cube. As at the beginning, 75% of the B' string is still sticking out of the top of the cube, 25% of the B' string is inside the cube, 75% of the C' string is still sticking out of the bottom of the cube, and 25% of the C' string is inside the cube.

If you've followed that, imagine the center of the cube shifts one-half inch upward along the B' string. In that position, 50% of the B' string is inside the cube, 50% is still sticking out the top, and 100% of the C' string is sticking out the bottom.

What of the other 4 strings?! They have moved along the B' string with the cube's center. That necessarily means that 50% of the A string is still inside the cube, 50% of the A string is still sticking out to the left of the cube, and 100% of the B string is still sticking out to the right of the cube. As in the prior circumstance,

13

50% of the A' string is inside of the cube, 50% is still sticking out to the rear, and 100% of the C string is sticking out of the cube's front and toward the viewer.

Leave the cube there for a while, and what we have is an act which can be described as: 50% of the A string + 50% of the A' string + 50% of the B' string. That's because 50% of each of those strings is *inside* the cube, and that's a way of *graphically* expressing the fact that, in such a circumstance, the actuality of the generator can be described as an act in which 50% A + 50% A' + 50 B' are actualized.

To top it all off, let's now imagine the center of the cube shifts another one inch leftward along the A string. In that position, none of the A string is sticking out to the left, 50% of the A string is sticking out to the right, and 100% of the B string is also sticking out to the right.

What of the other 4 strings?! They have moved along the A string with the cube's center. That necessarily means that, in this new current circumstance, 50% of the A' string is inside of the cube; 50% of the A' string is still sticking out to the rear; 100% of the C string is sticking out of the cube's front and toward the viewer; 50% of the B' string is inside of the cube; 50% of B' is still sticking out of the top; and 100% of the C' string is still sticking out of the cube's bottom. The result is an act we can describe as: 50% B' + 50 % A' + (100% A – 50% B). The "- 50% B" is merely a way of saying that 100% of the A string has been taken into the cube; but, half of that has been handed over to the B string.

Such, then, is how generators engage in locomotion in Haasian cosmology. Speaking *exactly*, however, the generators do not move from one *place in space* to another *physically* adjacent to it. On the contrary, they change from one particular *act* to another which is *logically* adjacent to the prior act.

H. LOCOMOTION IN A STRAIGHT LINE:

If you've followed that, you can now begin to understand how, in Haasian cosmology, a given *form's* center can move away from the centers of all other forms. All we need do is say this: Every form is concentric with some generator's actuality which then becomes its *carrying* generator. Naturally, the center of the form then goes wherever the center of its carrying generator's actuality goes. When the universe was created, the forms and the generators were created one at a time in order to give each carrying generator time to carry its "piggy-back" form's center away from the center of creation.

But, where is each carrying generator going? Earlier, we said that every form is in potency to something which is logically outside of the universe, and this something is what we can call a "heavenly sextet", since it presents each form with six frames of reference. There is one of these heavenly sextets for each form, and the logical distinction between each heavenly sextet is the form with which it is associated. The forms thus serve as the means by which the first heavenly sextet is able to clone itself.

A further kind of reproduction takes place among the heavenly sextets because of the logical distinction between a heavenly sextet *im*mediately associated with a given form, and a heavenly sextet *mediately* associated with that given form. The latter are called heavenly *benchmarks*. Heavenly sextets always rotate *rhythmically*, and benchmarks not always so.

But, to get back on the track, let's return to what's in potency to what. Forms are in potency to something heavenly; but, generators are in potency to the actuality of a form, and the actuality to which a generator is in potency is determined by the current OD of the form which it is carrying.

Remember what we said earlier about the difference between categorical, generic, and specific OD's. If a generator is carrying a categorical form, then it is in potency to the next highest categorical form; if a generator is carrying a generic form, then it is in potency to the next highest generic form; and, if a generator is carrying a specific form, then it is in potency to the next highest specific form.

For example, if a generator's piggy-back form is currently at its native categorical OD of 2^{128}, then the generator is in potency to the form whose current categorical OD is 2^{256}. Being in potency to such a form, the generator's potency can be described as an outer cube with a "diameter" equal to that of the universe and composed of 2^{768} (*i.e.:* 1.5×10^{231}) small cubes arranged in 2^{256} layers each of 2^{256} rows of 2^{256} cubes each. That, of course, means that each of the generator's six "strings of potency" would be trillions of light years in length.

For another example, if a generator's piggy-back form is currently at its native categorical OD of 2^{64}, then the generator is in potency to the form whose current categorical OD is 2^{128}. Being in potency to such a form, the generator's potency can be described as an outer cube with a "diameter" equal to that of 2^{-128} times that of the universe and composed of 2^{384} (*i.e.:* 3.94×10^{115}) small cubes arranged in 2^{128} layers each of 2^{128} rows of 2^{128} cubes each. Each of the generator's six "strings of potency" would be no longer than one-half the diameter of the hydrogen atom.

Obviously, the length of each generator's six strings of potency varies as the current ontological distance of the form to which it is in potency. So does the number of "locations" on each string. In Haasian cosmology, each of those "locations" has a length of roughly 10^{-47} cm., meaning: Whenever 4 of a given generator's strings of potency move along a fifth one, they instantaneously make a "quantum leap" of roughly 10^{-47} cm. at regular intervals. At least, that's the way creatures of sense would describe it. It would be more accurate to say that, at regular intervals determined by the current OD of the generator's piggy-back form, the generator shifts from its current act to one of the 6 (26 if diagonal shifts are possible) acts which are logically adjacent to its current act.

As you've perhaps guessed, the number of different acts which a generator can perform (*i.e.:* the number of smaller cubes comprising its "cube of potency") equals the cube of the current OD of the form to which it is in potency. Reading that, be sure to remember that the form to which a generator is in potency is not the same as the form it is carrying "piggyback".

I. CURVED LOCOMOTION:

In the above, there is no trace of *curved* motion. In Haasian cosmology, curved motion among the forms and generators is not possible except thru the influence of the heavenly sextets and benchmarks (This is a return to the pre-Galileo concept of a dichotomous universe in which the "heavenly" side is the origin of the "earthly" side's motion and is such by virtue of its motion in perfect circles.). How so?!

SUBSTRATA

Every form's heavenly sextet does a rhythmical "dance" in reference to its heavenly benchmark. It performs that heavenly "dance" in 3 planes. In the *primary* plane, the heavenly points A', B', C, and C' shift uniformly and rhythmically around the axis formed by A and B; in the *secondary* plane, A, C', B, and B' shift uniformly and rhythmically around A'C; and, in the *tertiary* plane, A, A', B, and C shift uniformly and rhythmically around B'C'.

How does that affect the forms? Recall the cube with the six strings protruding from it. The A string "points" to a heavenly A, the B string to a heavenly B, the A' string to a heavenly A' and so forth. If the heavenly points A', B', C, and C' shift uniformly in a rhythmical manner around the axis AB, then the strings pointing at them will follow, and the cube will rotate around the axis formed by the A and B strings. As the form rotates, so does the potentiality of the generators in potency to the form's actuality; and so, the motion of the generators becomes curved because, in effect, they move in straight lines but through curving space. In Haasian cosmology, space, even without time taken into consideration, is 12-dimensionally (*i.e.:* 4-dimensionally in each of 3 planes) dynamic rather than three-dimensionally static.

But, how do these 3 planes work together? Well, first of all the heavenly benchmarks increase in complexity. That's possible because, for shifts in the primary plane, the logical distinction of the shift from A' to B' is different from the logical distinction between the shift from A' to one-half the shift to B', and the logical distinction there is different from the logical distinction between the shift from A' to one-half of that one-half, and so forth.

To explain the final effect, take the heavenly sextet immediately associated with the form native to 2^{256}. It "dances" in reference to a heavenly benchmark which admits of 2^{128} (*i.e.:* the square root of 2^{256}) different frames of reference in the primary plane, 2^{128} different frames of reference in the secondary plane, and 2^{257} (*i.e.:* 2×2^{256}) different frames of reference in the tertiary plane, and there is a definite ratio between the "dances" performed in each of those planes.

Let's illustrate that with a form currently at its native OD of 2^8 (*i.e.:* 256). To do that, picture the cube with the six wires protruding from it. Grasp the two ends of the AB wires (axis of the primary plane) and make the cube and the other four wires rotate around it. However, do it in a definite number of stages. First, roll it a sixteenth (square root of 256) of a complete turn; pause; twirl it another sixteenth of a complete turn; pause; twirl it another sixteenth of a turn etc., until you complete a rotation. Now, grasp the two ends of the A'C wires (axis of the secondary plane), and make the cube and the other four wires rotate around it only a sixteenth of a complete turn. As you do so, imagine a sixteenth of a turn around the AB axis. Pause. Leave the A'C wires alone as you do another complete turn around AB by shifting a sixteenth of a turn sixteen times. When you've completed that second turn around AB, shift another sixteenth of a turn around A'C as you start your third turn around AB.

For every *shift* around A'C (secondary plane) there will be one *complete turn* around AB (primary plane) involving 16 shifts. When you finally complete a turn around A'C, there will have been 16 shifts in the secondary plane, and 16 complete *turns* around AB for a total of 256 *shifts* in the primary plane around AB.

Now, grasp the two ends of the B'C' wires (axis of the tertiary plane), and rotate the other 4 wires 1/512th of a turn around the B'C' wires. Pause. Repeat what we've done above before you roll the cube another 1/512th of a turn around B'C'.

When you get done with a complete turn around B'C' (tertiary plane), you will have done 8,192 complete turns in the primary plane (*i.e.:* 16 in each of 512 positions in the tertiary plane), 512 complete turns in the secondary plane (*i.e.:* 1 in each of 512 positions in the tertiary plane), and 1 complete turn in the tertiary plane. That will give us 131,072 shifts in the primary plane (*i.e.:* 16 in each of the 8,192 turns, and the total number of shifts will equal 2 times the square of the form's native OD), 8,192 shifts in the secondary plane (*i.e.:* 16 in each of the 512 turns, and the total number of shifts will equal 2 times the cube of the square root of the form's native OD), and 512 shifts in the tertiary plane.

The number of positions in the tertiary plane always equals 2 times the *current* OD of the form (In this case, 2 x 256 = 512.), and the number of positions in the primary and secondary planes equals, in each, the square root of the *native* OD of the form (if it has a square root, otherwise it's a bit more complicated), and that is how "nature" always "knows" the native OD of any given form.

J. GRAVITY, ELECTROMAGNETIC RADIATION, GALAXIES, AND ATOMS:

In Haasian cosmology, gravity is born when forms accelerate upward from their *native* to a new *current* OD. As they do so, they capture and make concentric to themselves forms *currently* at each of the OD's thru which they have passed plus—if the acceleration commences at or below the mirror threshold—forms currently at a definite number of the OD's below the native OD at which the acceleration commenced (That's because what happens *upward* on one side of the mirror is reflected *downward* on the other.). The result is super particles composed of an astronomical number of forms which have once again fused their centers together but remain logically distinct from one another because each is currently at a discrete OD.

If, after accelerating, a form decelerates, it releases one form for each OD it leaves behind. However, each form released is released at the OD of 2^{128}; and so, each released form's carrying generator at least tries to be in potency to the form of the universe (There's a very technical reason why it may not succeed—a reason having to do with the ability to shift the *center* of the generator's potency over vast distances instantaneously.) and to race for that form's outermost reaches at a velocity standard to all generators which are carrying forms currently at 2^{128}. That's Haasian cosmology's way of explaining electromagnetic radiation.

The most important acceleration takes place when a form native to 2^{128} accelerates to within a "stones' throw" of the form of the universe at 2^{256}. For, that acceleration brings about the formation of what Haasian cosmology calls "the 8 reverse categories", and these reverse categories account for the formation of the galaxies, clusters of galaxies, clusters of clusters, and so forth. How is that?

The **OD** at which each of the reverse categories commences is calculated so: (1) $2^{256} \div 2 = 2^{255}$; (2) $2^{256} \div 2^2 = 2^{254}$; (3) $2^{256} \div 2^4 = 2^{252}$; (4) $2^{256} \div 2^8 = 2^{248}$; (5) $2^{256} \div 2^{16} = 2^{240}$; (6) $2^{256} \div 2^{32} = 2^{224}$; (7) $2^{256} \div 2^{64} = 2^{192}$; (8) $2^{256} \div 2^{128} = 2^{128}$. As a result, there are six sectors of the universe and, in each, at least one "division" of galaxies whose leading form is currently at an **OD** somewhere between 2^{255} and 2^{256}, and that leading form is riding a generator in potency to, and moving toward the outer limit of, the actuality of the form of the universe. In that "division", there are several "orders" of galaxies each having a leading form which is currently at an **OD** somewhere between 2^{254} and 2^{255}, and each of those leading forms is riding a generator in potency to, and moving toward the outer limit of, the actuality of the leading form of the "division" to which the "order" belongs. In each of those "orders" of galaxies, there are several "classes" of galaxies each having a leading form which is currently at an **OD** somewhere between 2^{252} and 2^{254}, and each of those leading forms is riding a generator in potency to, and moving toward the outer limit of, the actuality of the leading form of the "order" to which the "class" belongs. In each of those "classes" of galaxies, there are several "families" of galaxies each having a leading form which is currently at an **OD** somewhere between 2^{248} and 2^{252}, and each of those leading forms is riding a generator in potency to, and moving toward the outer limit of, the actuality of the leading form of the "class" to which the "family" belongs. In each of those "families" of galaxies, there are several "super-clusters" of galaxies each having a leading form which is currently at an **OD** somewhere between 2^{240} and 2^{248}, and each of those leading forms is riding a generator in potency to, and moving toward the outer limit of, the actuality of the leading form of the "family" to which the "super-cluster" belongs, and so forth.

That "most important acceleration" also causes categorical forms at 2^1, 2^2, 2^4, 2^8, 2^{16}, 2^{32}, 2^{64}, and 2^{128} (an 8 way combo) to combine in an unusual fashion to form a microscopic version of the reverse categories, and the result is a cluster of super particles which then cause a form native to OD 1 to try to accelerate away from them. For, just as no acceleration can capture the *outermost* form *above* the "mirror threshold", so also can no acceleration capture the *innermost* form *below* the "mirror threshold". The form native to OD 1 now fuses with many of its own kind as it tries to escape capture and, thereby, produces a super particle in which all the components are concentric. It then orbits the 8 way combo at a current OD between 2^{126} and 2^{128}. There, it awaits every opportunity to complete its escape by shifting to a current OD of 2^{128}. With a shift to that current OD, it would then race away at the speed of light.

K. INTERGALACTIC SPACE TRAVEL:

Perhaps the most important effect of the 3 perpendicular planes is this: The velocity of all curved motion in the *secondary* plane is many times greater than it is in the *tertiary* plane, and the velocity of all curved motion in the *primary* plane is many times greater than it is in the *secondary* plane (In Haasian cosmology, the speed of light is the velocity limit for all *rectilinear* motion on the part of carrying generators operating outside the confines of the atom; but, it is by no means the velocity limit for *curved* motion on the part of the forms.). For observers in the tertiary plane (ourselves)

23

these very high speeds (as much as 2^{256} times the speed of light by *our* reckoning but not by the reckoning of those *in* those planes) in the secondary and primary planes are observed as curved magnetic fields. If observers are advanced enough in technology to utilize motions in the primary and secondary planes, then they can go anywhere in the universe in what we would call an astronomically small fraction of a second.

L. TIME

For forms and generators, time is relative. For the heavenly sextets in the primary plane, it is absolute. Why is that?

For forms and generators, time is the measure of the rate at which they change from act to act. But, as forms accelerate upward in current OD, the rate at which they change from act to act slows down. As a result, time, too, slows down for such forms. Conversely, as forms accelerate upward in current OD, the rate at which their carrying generators change from act to act increases. As a result, time, too, speeds up for such generators. In short, as ultimate constituents increase the velocity of their rectilinear motion, time slows down from the standpoint of their forms, but speeds up from the standpoint of the generators carrying their forms.

For heavenly sextets, time is the measure of the rate at which they change from heavenly benchmark to heavenly benchmark in the primary plane. Since that rate is forever the same for every member of every heavenly sextet, time is absolute from the standpoint of the heavenly sextets in the primary plane.

According to Haasian cosmology, every heavenly sextet member shifting in the primary plane forever shifts from one heavenly benchmark to another approximately every 7.2×10^{-96} seconds as we define seconds. In Haasian cosmology, that fraction of an earthly second becomes the definition of what is called "one alphakronon". As a rule, then, an earthly second is roughly 1.4×10^{95} alphakronons of absolute time.

In Haasian cosmology, the universe is currently in its ninth epoch, and the absolute duration of that epoch will be 2^{385} (*i.e.:* $7.8804012 \times 10^{115}$) alphakronons, which is to say approximately 5.7×10^{20} seconds or approximately 1.8×10^{13} Earth years. The number 2^{385} is 2 times the cube of the square root of the OD of the form of the universe.

At the end of 2^{385} alphakronons, the 9th epoch of the universe will come to an end, and the 10th epoch will commence. In that 10th epoch, the OD of the form of the universe will be $(2^{256})^2$, the number of forms created native to OD 1 will be 6^{512}, and the duration of the 10th epoch will be $2[(2^{256})^3]$ alphakronons.

Finally, there is subjective time, which is to say time as we *experience* it. In Haasian cosmology, consciousness is a series of individual acts on the part of a special kind of form called a "soul". Each of a given soul's acts of awareness has a definite duration determined by the current ontological distance of the soul. *Increase* a soul's current OD, and the duration of each of its acts of awareness *increases*. *Decrease* a soul's current OD, and the duration of each of its acts *decreases*. As long as the soul is performing the same, one act, it undergoes no kind of internal change whatsoever and, therefore, has no way to *experience* the duration of its acts. As a result, a soul experiences each of its acts of awareness as having zero duration; and yet, Haasian cosmology says that, for souls at our OD of approximately 2^{155}, each act of awareness has a duration of roughly 3.5×10^{93} alphakronons. Increase a soul's OD to

2^{200}, and it would seem to that soul that the Earth is whirling around the sun faster the tip of a airplane's propeller blade whirls around its shaft. Decrease a soul's OD to 1 (a reduction which would put it at the innermost depths of an electron), and it would seem to that soul that photons hardly ever move at all. As far as *experience* is concerned, what face the world around us shall present to a given observer is purely relative to the current OD of that observer's soul.

Such, then, are the basics extracted by 40 plus years as a philosopher inwardly reflecting upon—on one hand—the tri-partite structure absolutely necessary to every act of consciousness. I say "on one hand", because a second target of introspection was heavily involved: *Something* in me is instinctively *addicted* to the notion that time, space, spatial things, and locomotion are all infinitely divisible. But, even from my childhood days—as early as the second grade—my *intellect* regarded this notion of "infinitely divisible" as self-contradictory gibberish more obviously ludicrous than is a square circle. Indeed, my intellect proclaims that notion so insane, I've often wondered how any but the most diabolical of intellectual frauds could cling to it for more than a micro-second. Consequently, hand in hand with my reflection upon what is absolutely *a-priori* to every act of consciousness, there was also an effort to figure out how time, space, spatial objects, and the movements of the latter thru the other two could strike our senses as being *continuous* when, as I saw it, they could not *possibly* be anything but <u>dis</u>continuous.

26

Preface To The Edition Of 1983:

(A)
PHILOSOPHY VS. SCIENCE:

i.
DID PHILOSOPHY SEEK WHAT ASTRO- & NUCLEAR PHYSICS DO?

But, they, owing to their love for their principles, fall into the attitude of men who undertake the defence of a position in argument. In the confidence that their principles are true they are ready to accept any consequence of their application. As though some principles did not require to be judged from their results, and particularly from their final issue! And that issue, which in the case of productive knowledge is the product, in the knowledge of nature is the unimpeachable evidence of the senses as to each fact.[1]

——ARISTOTLE: *On The Heavens*; Book III; Chap. 7; 306a: 13-17. *Great Books Of The Western World*; Vol. 8; pg. 397 bottom left and top right.

Lack of experience diminishes our power of taking a comprehensive view of the admitted facts. Hence those who dwell in intimate association with nature and its phenomena grow more and more able to formulate, as the foundations of their theories, principles such as to admit of a wide and coherent development: while those whom devotion to abstract discussions has rendered unobservant of the facts are too ready to dogmatize on the basis of a few observations. The rival treatments of the subject now before us will serve to illustrate how great is the difference between a 'scientific' and a 'dialectical' method of inquiry.[2]

---

[1] **NOTE OF APRIL 20, 2002:** In other words, one's principles should always allow one to point one's finger at something in the world around us and say: "See that there. That's what my words mean." Is Aristotle a *philosopher* or a *scientist*?! But, if you think that's something watch what he says in the next quote.

[2] **NOTE OF APRIL 20, 2002:** In other words, the "dialectical method" spends too much time spinning abstract arguments and not enough time observing nature and, as a result, goes

———ARISTOTLE: *On Generation And Corruption*; Book I; Chap. 2; 316a: 5-13. *Great Books*; Vol. 8; pg. 411 upper right.

That philosophy is a deductive method of inquiry, I dare say no man will deny. On the other hand, is philosophy an inquiry into nature? That is to say, is philosophy an attempt to deduce such things as the diameter of the universe or the properties of the ultimate sub-atomic particles? Ever since the seventeenth century, when science dethroned philosophy, it has been fashionable for philosophers to insist that philosophy is *not* an inquiry into nature. For example, Jacques Maritain, the great French philosopher, writes:

> It is also very true that metaphysics is of no use in furthering output of experimental science. Discoveries and inventions in the land of phenomena? It can boast of none. Its heuristic value, as they say, is absolutely nil in that area.
> ———*The Degrees Of Knowledge*; Translated from the French by Gerald B. Phelan; Charles Scribner's Sons, New York; pgs. 3-4

---

nowhere; conversely, the "scientific method" spends so much time and effort observing nature it can be said to live in "intimate association" with nature and, as a result, ever moves ahead. This is a *philosopher* talking?! Isn't this the favorite of science's *current* complaints against philosophy? Incidentally, Aristotle elsewhere gives "dialectical method" a different meaning.

> The ancients thought their philosophy of
> nature was a science of the phenomena of
> nature. That was their misfortune.
> ——*Op. Cit.* pg. 175

Yes! By Monsieur Maritain's own admission, the
ancients thought that philosophy is indeed a deductive
method of inquiry into the "phenomena of nature".
That's why even Aristotle himself—one of the three
greatest philosophers in history—wrote much about
whether or not the world is a sphere, what its diameter
might be, how large it might be in comparison to the
stars, and whether it rotates around the sun or the sun
around it.

> Most people—all, in fact, who regard the
> whole heaven as finite—say it lies at the
> centre. But the Italian philosophers
> known as Pythagoreans take the contrary
> view. At the centre, they say, is fire, and
> the earth is one of the stars, creating
> night and day by its circular motion
> about the centre.
> ——***On The Heavens***; Book II; Chap. 13;
> 293a: 19-23. ***Great Books***; Vol. 8; pg. 384
> lower right.

> This indicates not only that the earth's
> mass is spherical in shape, but also that as
> compared with the stars it is not of great
> size.
> ——***On The Heavens***; Book II; Chap. 14;
> 298a: 18-20. ***Great Books;*** Vol. 8; pg. 389
> lower right.

ii.
HUMILIATED, PHILOSOPHY
RETREATS INTO MUMBO-JUMBO:

What changed all of that? How did it come about that philosophers were suddenly no longer also and automatically scientists theorizing about the heavens?

In the seventeenth century, science used the microscope and the telescope to prove most dramatically that, **if** philosophy was indeed a deductive method of inquiry into the phenomena of nature, then it was an inquiry so pathetically mistaken in its conclusions as to stagger the mind of man. As a deductive method of inquiry into the phenomena of nature, philosophy was suddenly the laughing stock of virtually every nook and cranny of the educated world. Overnight, the name "philosopher" suddenly meant, *at best*, the most ignorant, thick-headed bigot the mind of man could conceive—*at worst*, a malicious sycophant dedicated to perpetuating human ignorance in an attempt to keep the masses in abject subservience to the absolute monarchs of Europe.

Predictably, those who still cherished the title of "philosopher" took the only retreat open to them as human beings afflicted with egos: Rather than admit they simply didn't know enough about philosophy to make the right deductions about the phenomena of nature, they insisted philosophy was *not* an inquiry into the phenomena of nature. In other words, they abandoned their faith in the wide ranging abilities of the deductive method of inquiry in order to preserve their pride in their own skill in the art of deduction, and such they accomplished by falling back upon nothing but

31

their most basic deductions and then insisting that those deductions were unaffected by the discoveries of science, since, according to the philosophers, those *fundamental* deductions pertained to something beyond the veil of the phenomena of nature. To put it still another way, they preferred to eviscerate philosophy rather than to abandon the egotistical illusion that—in at least *some* of their philosophical concepts—they had achieved as great a perfection and refinement as it is possible to attain.

With philosophy excluded from the phenomena of nature, words such as "substance", "substrata", "space", and "matter" no longer have, and no longer *need* to have, any *concrete* meaning. That is to say, they no longer apply to anything given to human experience—to anything to which we can point and say: "That's what the word in question indicates." Instead, they refer only to a *notion*—to an esoteric enigma which is a "necessary notion" only "indirectly known in an analogous way in virtue of something else". As Father Joseph Owens, C. Ss. R., puts it:

> But the change of the one thing into the other requires a common subject, according to the very notion of change. Such a common subject will be unobservable both in principle and in fact, because it is what loses and acquires the most basic of forms and so of itself has not even the most rudimentary principle of knowability or observability. It has to be known in virtue of something else. That "something else," quite naturally, will be the observable subject in accidental change, like the wood that becomes a bed or the bronze that becomes a statue.

Some corresponding subject has to be present for substantial change. In that analogous way, then, the subject of substantial change, namely matter, is indirectly known. It is known as the conclusion of scientific reasoning in the Aristotelian sense of 'scientific'.
——*Matter And Predication In Aristotle. Aristotle; The Collected Papers Of Joseph Owens*, Edited By John R. Catan; State University of New York Press; Albany, © 1981; pg. 45.

The Aristotelian matter has not been superseded nor even touched by the stupendous progress of modern physics. Nothing that is measurable can perform its function in explaining the nature of sensible things, and by the same token it cannot be brought forward to account for anything that requires explanation in measurable terms. Any type of matter dealt with by chemistry or modern physics would in comparison be secondary matter, and not matter that is a principle in the category of substance.
——*Op. Cit.* pg. 47

Understandably, such an interpretation of how philosophical terms have meaning readily leads to the conclusion that philosophy is a kind of mystical experience given only to the spiritual elite. It's a grace from God which lifts them to a lofty and most extraordinary level of awareness—a level at which the intellect is suddenly able to grasp things so "spiritual" that they cannot possibly be pinpointed either with the finger or any of the exact mathematical formulae of physics and chemistry. It comes as no surprise, then, to find Jacques Maritain writing, in **The Degrees Of Knowledge**, as follows:

> How is it possible to speculate about geometry in space if figures are not seen in space? How is it possible to discourse on metaphysics if quiddities are not seen in the intelligible? . . . A Jesuit friend of mine claims that since the fall of Adam, man has become so ill suited to understanding that the intellectual perception of being must be looked upon as a mystical gift, a supernatural gift granted only to a few privileged persons. That is obviously a pious exaggeration. Yet, it does remain true that this intuition is, as far as we are concerned, an awakening from our dreams, a step quickly taken out of slumber and its starried streams. For man has many sleeps. . . . There is a sort of grace in the natural order presiding over the birth of a metaphysician just as there is over the birth of a poet.
> ——*Op. Cit.* as translated from the French by Gerald B. Phelan. Charles Scribner's Sons; New York, pg. 2.

Alas for us the "non-poets" still lost in the "starried streams" of intellectual slumber! If we fail to see the importance of words which refer to nothing more than "a necessary *notion*" which is only "indirectly known in an analogous way in virtue of something else", then it is only because—unlike the mystically re-born "philosophers"—we've not yet received that "sort of grace in the natural order" which presides over the second birth. We've not yet been awakened from our dreams and blessed with the capacity to lift our wide-awake eyes heavenward where the reborn see heavenly, mystical, and poetic things. Yes! We are still pigs with our snouts contentedly buried in the slime of the earth; consequently, for us—because we *are* slime eating, ignorant pigs content to wallow in our own ex-crement—everything must be either translated into something as concrete, fixed, and real as a sense image, or else it is meaningless to us. One can only marvel at the patience and fortitude which allows the "philoso-phers" to bear the unbelievable burden of spending a lifetime in the midst of such pathetically inferior, fellow human beings as we are.

The hidden purpose behind such pomposity is easily detected. It is an attempt to produce a purely dialectical method of inquiry which, because it restricts itself wholly and entirely to *abstract* discussions, pro-duces principles which can never be judged by refer-ence to the "unimpeachable evidence of the senses as to each fact" (*cf.:* the quotes from Aristotle at the opening of this preface). Naturally, once you do that, you can rest assured that no one will ever again embarrass you as totally as science once embarrassed you, because no one will ever again succeed in pinpointing the exact meaning of what you're saying.

To say the above another way, what we have here is a small group of men who were once dragged before the court of empirical evidences and there de-feated as totally as any intelligent being *can* be defeated. Now, to guarantee themselves that they will never again be *that* totally routed, they have *predictably* sought to find some way to "prove" that they are *intrinsically beyond* the jurisdiction of that court. Necessarily, therefore, they have turned to terms, definitions, and principles whose meaning, *they propose*, cannot be judged by "the unimpeachable evidence of the senses".

Unfortunately, the so-called "philosophers" merely delude themselves. They *vainly* imagine that their ingenious little charade has genuinely warded off the thrusts of the scientific world—that they have proven that science's achievements nowhere touch upon, or in any way affect, those concepts which *legitimately* belong to philosophy. In fact, the "philosophers" have only confirmed the most telling of the charges hurled at philosophy by science, namely: that the concepts of philosophy are nothing but words with no non-verbal meaning whatsoever and, as such, are concepts whose exact meaning, relevance, and importance to mankind can *never* be *definitely* established. The result is that, by retreating from the court of empirical evidences and falling back upon "indirect knowledge drawn from something else by means of an analogy", the so-called "philosophers" have done nothing except to justify the assertion that what they *call* "philosophy" is nothing more than a logomachy—a "word-war" which can never serve *any* purpose whatsoever save to afford the lovers of argumentation an arena in which they can go on arguing forever over every jot and tittle in the system without ever proving anything one way or another. In short, in their attempts to preserve their egos and to maintain some sense of *deserved* influence in the intellectual community around them, those who are commonly called "philosophers" have done no more than to lay down lame excuses which have *thoroughly justified* the almost *universal* rejection with which they have been greeted by the scientific community.

iii.
THE MUMBO-JUMBO'S NEGATIVE VALIDITY:

Having said that, though, one must hasten to add the other side of the story: Though the absurd claims of the "philosophers" are indeed nothing but infantile rationalizations *justifying* the reaction of the scientific world, it still remains true that the reaction of the scientific world is an <u>over</u>-reaction. It is an over-reaction because, contrary to what the scientists would like to believe, the concepts of philosophy *do* have *some* validity. Unlike what the "philosophers" would like to believe, philosophy's concepts do *not yet* have any *positive* validity; nevertheless, unlike what the scientists would like to believe, they *do* have what I would chance to call *negative* validity. What does that mean? Let's explain it by contrasting what I mean by positive and negative validity.

To have a *positively* valid concept of God is to be face to face with God. To have a *positively* valid concept of the color red is to see the color red. To have a *positively* valid concept of the sound of a trumpet is to hear a trumpet and so forth. A man unable to have a visual sense image cannot have a *positively* valid concept of anything *seeable*. A man without a sense of touch cannot have a *positively* valid concept of anything *touchable*. An intelligent being, *strictly limited* to those things which can be *sensed, cannot* have a *positively* valid concept of anything whose internal characteristics are encountered *only* in a form of experience *other than* sense-experience. In short, to have a *positively* valid concept of any particular thing is to *encounter* (*i.e.:* behold "face to face") the *internal* characteristics of that particular thing.

Suppose, now, I tell you the exact location of a buried treasure. You know exactly *where* it is, but I tell you *nothing* of what it looks like, or smells like, or feels like, or tastes like, or sounds like. Indeed, it may well be that that treasure cannot be *sensed*. Perhaps, when you go to the spot in which that treasure is located, you are accelerated to infinite velocity and suddenly become capable of a new form of experience which is wholly and entirely unlike sense experience. Perhaps only then do you become capable of beholding that buried treasure's internal characteristics. Until that "boost to infinity" occurs, what *do* you *know* of that buried treasure? Clearly, you know nothing of its *internal* characteristics; you know nothing of what it "looks" like. All you know is *where* it should be found, **if** and *when* and it *is* found. You also know that to find it is to be affected in a particular manner, and thus you know something of its effects upon others; nevertheless, you can form no "picture" or "image" of what characteristics will be pressed upon your consciousness *if* and *when* you come "face to face" with the treasure.

Therefore, as a *positively* valid concept, your knowledge of the buried treasure is a *blank*, which is to say the buried treasure is a *positively unknown nullity*— a pure question mark which gives not a single, solitary clue to even so much as *one* of the internal characteristics of whatever might someday fill that blank. Nevertheless, it is *indirectly* and *negatively* known in that, *by means of our positively valid concepts of the world around us*, we are able to *relate* the positively unknown nullity to what is positively known, which is to say, we *indirectly* know at least *one* of the *external* characteristics of that mysterious blank. In short, to have a *nega-*

tively valid concept of any particular thing is to encounter, *in another thing*, a positively valid concept of the *external* characteristics of the first thing.

Hearing the above, some will promptly remark: "Aren't you merely parroting what you quoted Fr. Owens saying in his description of 'Aristotelian matter'?" To that, I will answer in the affirmative but hasten to add three major qualifications: (1) The vast majority of those who are called "philosophers" would apparently have us believe that philosophy is *limited* to what we've just described as "negatively valid concepts"; whereas, I say philosophy *does*, in certain cases (such as "pure potency", as we shall see), lead to what I call "positively valid concepts". (2) Though they admit that their philosophical concepts give them only *indirect* knowledge *in an analogical way by virtue of something else*, the vast majority of those who are called "philosophers" seem to think that—notwithstanding that "indirectness"—the concepts of philosophy are somehow *positive* and somehow reveal the indirectly known thing *itself* to the gaze of the intellect; whereas, I say that all such "negatively valid concepts" (as I choose to call them) are *positively blank* concepts which *in no way whatsoever* reveal *anything* of the indirectly known thing *itself* to the gaze of the intellect, which is the same as saying "indirect concepts" (as others might call them) in no way reveal to us anything of the *internality* of the *indirectly* known thing. (3) The vast majority of those who are called "philosophers" refuse to admit how terribly far the "negatively valid concepts of philosophy" *currently* are from being as *adequate* as they could be; whereas, I freely admit that, till now, philosophy's concepts have been so far from being as *adequate* as they *could* and *should* be, that, as a result, philosophy presents us with so little information that philosophy is *manifestly* not even remotely worth a second of any thinking individual's time. What does that mean? Read on!

What is the value of a negatively valid concept? Negatively valid concepts serve as signposts telling us where something is missing. They are signals warning us where our positively valid concepts fall short of a full explanation of what is positively known. Though they reveal *nothing* of the *internal* characteristics of what they tell us is missing, they, nevertheless, highlight the deficiencies and weak points in what we *do* know *positively*. The problem is, though, that not all negatively valid concepts accomplish that objective with the same degree of thoroughness. Why do I say that? Let's examine that question.

Every *positively* valid concept is—*by its very nature—adequate*. By that, I mean this: There is, for example, no such thing as an *inadequate* experience of the color red. Either you experience the color red, or you don't. If you *do* experience it, then you immediately see the color red for exactly what it is. There is no such thing as another individual who sees, in his experience of the color red, a more revealing experience of what red is *in itself* than you do.

That is far from being true in the case of *negatively* valid concepts. *Negatively* valid concepts admit of differing degrees of adequacy. By that, I mean some reveal more than others concerning the *external* characteristics of the *positively unknown*. For example, suppose I say to you that there is a buried treasure somewhere in the universe. Since I've given you no positive knowledge whatsoever concerning the internal characteristics of that buried treasure, I have given you only a *negatively* valid concept of the buried treasure. Still, what I have told you is so general and vague that you would be a fool to make even the slightest effort either to confirm or to refute my statement. Knowing that you

40

can't possibly check out every possible location everywhere in the universe, you would rightly dismiss my assertion as a "truism" with which only fools would concern themselves. On the other hand, suppose you and I are standing in a field and I point to the ground directly beneath our feet and swear there is, two feet beneath the surface, a buried treasure worth all the money in the world. That is a negatively valid concept which you could either *empirically confirm* or *empirically disprove* with very little effort. It, therefore, is an *adequate* negatively valid concept.

Therefore:

(1) An *adequate* negatively valid concept is one which conveys to the beholder enough information to convince the beholder that he can and ought to attempt either the *empirical confirmation* or the *empirical refutation* of that negatively valid concept;

(2) An *inadequate* negatively valid concept is one which conveys to the beholder so little information as to convince the beholder that he neither can nor should attempt either the *empirical confirmation* or the *empirical refutation* of that negatively valid concept.

Quite possibly, one might induce most philosophers—so-called or otherwise—to agree that philosophy deals with negatively valid concepts. Most of those who are *called* "philosophers", however, would no doubt loudly denounce *my* distinction between adequate and inadequate negatively valid concepts. They would insist, instead, that philosophy deals only with concepts which can *never* be hauled before the court of empirical evidences either for the purpose of *confirming* them or for

41

the purpose of *refuting* them. I, of course, say that they will say that because they *must* say that in order to preserve their own egos. They know only too well that, *up to this time*, the concepts of philosophy have no capacity whatsoever to earn even the *attentions* of the scientific community, let alone empirical confirmation by that community. As a result, they know only too well that, **if** their ability to maintain their influence within the whole intellectual community must rely upon their ability to make the concepts of philosophy verifiable by science, then all influence is hopelessly lost. A lifelong relegation to the ranks of the ignorant nobodies whose intellectual efforts draw nothing but jeers from those whom the public *calls* "the experts"—that is something which virtually none of the so-called "philosophers" can endure.

True philosophy and *true philosophers* are ever sustained throughout their entire being by the free gift, *from* God, of faith *in* God. By that free gift of God's grace, the very tips of their bodies' nerves *feel* a marvelous confidence that, *no matter how totally they may err and miss the mark in this life, still* God will not leave them naked and embarrassed *on judgment day*. With the free gift of God's grace rendering their egos impervious to the shame that normally comes from total refutation before the court of empirical evidences, *true* philosophy and *true* philosophers hesitate not to submit their thoughts to the court of empirical evidences and to say: "Here! Totally refute my conclusions. Prove utterly that I am *still* nothing but the most stupid of fools whose thoughts are yet mindless gibberish unfit for anything but the halls of an insane asylum. What do I care what *you* say, O mere mortals? By the free gift of God's grace, I am *utterly undaunted* in my confidence that I *am on* the right path and *will eventually* find the truths which, as I myself must admit, you have proven most conclusively are not *currently* in my grasp."

False philosophy, though, and *false philosophers* have not the God-given power to trust in *future* justification on Judgment Day. As a result, their egos cry out to be proven correct and fruitful *here and now*. But, they *know* that—if they must submit their deductions to the court of empirical evidences for confirmation by that court—either they will be proven to be utterly mistaken, or it will be shown that they are yet men whose deductions are still so inadequate, barren, and impoverished as to render them unworthy of any attention or influence whatsoever. That, though, would *justly* precipitate rejection by every thinking individual and leave the "philosophers" excluded from an intellectual community totally indifferent to them. Therefore, for the sake of preserving their egos and some sense of importance and influence in the intellectual community, they *must* take refuge in a lie and insist that *all genuine* philosophy *always* arrives *only* at principles, terms, and definitions which can *never* be either verified or *disproved* by empirical evidences.

iv.
3 VIEWS ON PHILOSOPHY'S FUTURE:

Lest anyone mistake what I am saying, let us digress here for a moment. There may be some "philosophers" blind enough to assert that I am here spouting the stale doctrine of empiricism—a doctrine which alleges that experience is the sole source of knowledge and denies there are universal and necessary truths which hold true regardless of past, present, or future experience. If any wish to accuse me of that, then let them know how mistaken they are. Quite the contrary to what *they* say, I do most emphatically assert that there *are indeed* universal and necessary truths which can be deduced from nothing more than mere reflection upon the nature of an act of consciousness.

The question at hand here, therefore, is *not* whether or not there is such a thing as universal and necessary truths unaffected by the data of the senses; the question is, rather, how far can philosophy go in developing the *content* of those universal and necessary truths. Can it possibly be that the deductive method of inquiry can so refine the universal and necessary truths of philosophy as to produce *adequate* concepts—concepts so rich in content as to win the admiration and approval of the court of empirical evidences? I answer that it most certainly *is* possible. Indeed, I go so far as to say that philosophy can *and will* eventually deduce *all* the laws of physics and once again throw open to mankind the possibility of re-entering the garden of Eden.[3] To accomplish such, all that's required is the grace of God through His only begotten Son, our Lord and Savior Jesus Christ, Who alone is true God *and* true man.

To express, in as clear a fashion as possible, the difference between empiricism, pseudo-philosophy, and true philosophy, one can draw a simple little picture. For the empiricists (*i.e.:* the scientists), all philosophy in general (and Aristotelianism in particular) is a useless weed which needs to be eradicated and incinerated

---

[3] **NOTE OF APRIL 6, 2002:** Even when I originally wrote the above sentence, the phrase, "re-entering the garden of Eden", did not even begin to mean to me a return to *grace*. It meant then and means now only that technology can advance to the point at which it can eradicate all human misery (including the debility of old age) and build space ships capable of intergalactic travel in what we would call a fraction of a second.

in a blast furnace. For the pseudo-philosophers, philosophy is a marvelously fruitful tree which has already borne virtually all the fruit it can ever possibly bear. This fruit, though, is purely *ethereal*. There is nothing whatsoever of the *material* in it, and, consequently, it can never in any way whatsoever be brought before the court of empirical evidences. For the true philosopher, though, philosophy is a tree which is utterly barren and, for the most part, dead. What little life it has, it has by virtue of nothing but the grace of God through Jesus Christ. To date, that grace has—for reasons discussed in another book—enabled that wretched tree to do no more than to put forth a few, pathetically fragile leaves. At the same time, though, that tree's *future* is—by the grace of God through Jesus Christ—rich with promise, and, one day, by that grace, that tree will suddenly blossom forth and bear such fruit as will astound the world. In that day, the concepts of philosophy will stand before the court of empirical evidences, and the scientists will all fall back in disbelief, collapse to the ground, and stammer in bewilderment: "Is not this the plant which we contemptuously dismissed as a worthless weed? How now is it become the mightiest of all trees—yielding in superabundance all the fruits which we had sought in vain from another and exceedingly different tree?"

Yes! For now, all the concepts of philosophy are *inadequate*. For now, philosophy has—*in and of itself!*—not a single, solitary shred of thought to merit that thinking men should devote even a second of their time to philosophy's *intrinsically* worthless concepts. *In and of itself*, philosophy is *most manifestly* nothing but a consummate fool's errand. But, there is more to philosophy than its *intrinsic* merits. It has been chosen by God, and God has borne witness to that fact by causing the Roman Catholic Church to honor philosophy above science. For that reason, men of *faith*—trusting in the unfailing guidance of the Catholic Church—have not hesitated to dedicate their entire lives to digging and scratching in what all the publicly acclaimed "experts" say is a worthless field. What do they care that mere mortals, called "experts" by other mere mortals, should despise them throughout the course of their lives? No matter how great the number of "experts" hurled against them by mere mortals; no matter how many the centuries they must labor without achieving anything but ridicule at the hands of those called "experts" by other men—the men of faith *know* what philosophy will one day achieve, and they will not be turned aside from their dedication.

V.
NOT VS. NOT YET:

A few pages earlier, Father Owens told us that "the Aristotelian matter" is "unobservable both in principle and in fact", that it "has not even the most rudimentary principle of knowability or observability". He told us it is "indirectly known" and "has to be known in virtue of something else" and "in that analogous way". He told us that "Nothing that is measurable can perform its function in explaining the nature of sensible things." He is completely mistaken. It is by no means true that nothing measur*able* can perform its function in explaining the nature of sensible things. It is true *only* that nothing measur*ed* can perform its function, and that is true only because science has not yet achieved the capacity to measure all that is there in the physical world waiting to be measured. It is by no means true that it *has* to be indirectly known in an analogous way in virtue of something else. It is only true that it *currently* has to be known in that way, because, till now, the Aristotelian concept of matter is *inadequately* developed and defined. It is by no means true that it has not even the most rudimentary principle of knowability or observability. It is true only that it *currently* has not the most rudimentary principle of knowability or observability, and that manages to be true only because, till now, the Aristotelian concept of matter is *inadequately* developed and defined. Finally, it is by no means true that the Aristotelian matter is unobserv*able* both in principle and in fact; it is only true that it is *currently* unobserv*ed* both in principle and in fact.

All that we *fully* learn from Aristotle, Father Owens, and company is that neither they nor any of their scientific opponents yet have *adequate* knowledge concerning what is meant by such terms as "matter", "space", "substance", and "substratum". In the final analysis, all this talk about "indirect knowledge in an analogous way by means of something else" is merely an inordinately proud, slippery eel's clever way of avoiding the embarrassment of having to admit that he simply does not yet have *adequate* knowledge of what he's talking about. By way of comparison, one might imagine a man blind from birth, who, because he has an *intense* inferiority complex, cannot admit he is blind. Suddenly, he comes upon an elephant. He wants to tell you about it. Faced with an object that huge, how shall he tell you about it without clearly and forcefully *demonstrating*, both to you *and* to himself, that he *is* blind? Were he not an unduly egotistical man, he would say to you: "I *know* that I have *truly* come into contact with something which can and ought to be communicated to you. Unfortunately, the object I need to describe to you is so immense—and my defective consciousness's grasp of it is so primitive, patchy, and inadequate—that I

47

simply do not know enough about it to be able to communicate it to you with any appreciable degree of effec-
tiveness." But, because he *is* an inordinately proud man,
he will *not* confess that the *main* problem is the flimsi-
ness of his knowledge of his subject; consequently, the
first thing he sets out to do is to convince you that the
main problem behind his attempts to tell you about the
elephant is that, first—*unlike all other* knowable
things—the elephant is a *poetic* entity which only the intellectually gifted ones can grasp; and, secondly, unlike
himself, you're obviously not yet one of the intellectually gifted ones.

When all the "face-saving" camouflage is cleared
away, though, we see what is *actually* happening: The
negatively valid concepts of the philosopher give him
just enough *inadequate* knowledge to know that—in our
explanations of the phenomena of nature—*something
very critically important is missing*. Nevertheless, when it
comes time to tell us what that missing element is, his
negatively valid concepts of that missing element are *so
inadequate*, he cannot even begin to give us even the
slightest clue to where we might look in the vastness of
the cosmos for that missing element, which is to say he
can tell us virtually nothing of even its *external* characteristics, let alone its *internal* ones. Like the man with
the buried treasure somewhere in the universe, he
leaves us with the need to search for a needle in a hay
stack trillions of light years in diameter. But, rather
than admit he has done that—and thus be forced to face
the *practical* worthlessness of the information he has
given us—he hands us instead (to paraphrase Aristotle)
the long rigmarole about principles which do not require to be judged from the unimpeachable evidence of
the senses as to each fact.

vi.
HOW ARISTOTLE SET PHILOSOPHY UP FOR ITS BIG FALL:

Since this whole mess started with Aristotle, let's ask how it *did* start with him? As he himself told us in the quotes given at the beginning of this preface, some are "too ready to dogmatize on the basis of a few observations". To paraphrase something else he said in those quotes, if we are to formulate, as the foundations of our theories, principles which admit of a wide and coherent development, then we must dwell in intimate association with nature and its phenomena. Sad to say, for the fallen sons of fallen man, to do that without the aid of microscopes and telescopes is a virtual impossibility. That's why, prior to the scientific revolution of the seventeenth century, there *were not* and *could not* have been any philosophical principles based on *adequate* observations. All that came before the great observations of the microscope and the telescope was nothing but "abstract discussions" which cannot be "judged from their final issue", which is to say all was *dialectics* which could never produce anything capable of having merit in the eyes of the court of empirical evidences.

Armed with what was, in his day, *necessarily* "a few observations", Aristotle defined "substratum" *and* "substance" (*i.e.:* substance *taken as a category*) in an erroneous fashion. He defined *both* as "that which is *neither* present in *nor* predicable of another". There is no reason whatsoever to *compel* us to define "substratum" in that fashion, and modern day observations of the phenomena of nature more than suggest to us that it is better to say that *some* substrata are definable as:

"that which *is present in* another but *not predicable of* that other".[4]

---

[4] **NOTE OF APRIL 6, 2002:** Perhaps the chief fountain of all the problems in philosophy comes from the phrase "present in". When either Aristotle or I myself speak of a first something (Let's call it "Alpha" for convenience's sake.) as being "present in" a second something (Let's call it "Beta" for convenience's sake.), we mean that Alpha is incapable of existence *apart from* Beta. There's where an ocean of confusion begins. For, the phrase "apart from" can be taken in two radically different ways. **ON THE ONE HAND**, "apart from" can mean that Alpha cannot exist, unless Alpha is *intimately bonded* to Beta. For example, this or that particular *shape* cannot exist unless some kind of "stuff" takes on that shape. We express that by saying that no given shape can exist *apart from* being *intimately bonded* to some kind of substratum (*i.e.:* "shapeable" "stuff") assuming the given shape. **ON THE OTHER HAND**, "apart from" can mean that Alpha cannot exist, unless Beta is also existing and exerting its influence on Alpha. For example, none of the objects we observe on a daily basis (whether people, animals, plants, steel girders, cars, or what have you) can exist unless temperature, air pressure, air content, etc., remain within particular ranges. Where "apart from" is taken in that sense, *intimate bonding* between Alpha and Beta is to no extent implied. From the day I first read Aristotle's use of "present in" and "incapable of existence apart from", I never understood *him* to mean anything other than that Alpha is incapable of existence apart from *intimate bonding* to Beta. More specifically, I never understood *him* to be referring to anything other than the obvious relationship between *characteristics* and *whatever exhibits* those characteristics. Since the manifest relationship between characteristics and what _has_ characteristics is one of *intimate bonding* of the former to the latter *or else*, I never understood Aristotle to be referring to anything other than *that* particular and very obvious manner of being "present in" another. To me, that was _so_ obviously what Aristotle meant, that, for years, I never had the slightest inkling that others might understand him in

The instant we make that distinction, it becomes possible to take the *class* called "substrata" (*i.e.:* "that of which everything else is predicable while it is not itself predicable of anything else") and to divide it into *two genera* of substrata, namely:

(1) **SUBSTANTIAL** substrata (*i.e.:* in the <u>*class*</u> "substrata", that <u>*genus*</u> which is defined as: that which is *neither* present in *nor* predicable of anything else);

AND

(2) **INSUBSTANTIAL** substrata (*i.e.:* in the <u>*class*</u> "substrata", that <u>*genus*</u> which is defined as: that which is *present in* a substantial substratum but is *not predicable of* that substantial substratum).

That, in turn, allows us to distinguish *three species* of substrata, namely:

(1) *Actually* substantial substrata (*i.e.:* that the internal characteristics of which, at this stage in the development of the universe, would be encoun-

---

even so little as a *slightly* different way, let alone a *radically* different one. For years, then, it never dawned on me to any extent whatsoever that there was any need to explain that "apart from" another means "apart from intimate bonding to" another; and so, for years, I never bothered to make that qualification. That, of course, is why the qualification never occurs in my work **Aristotle's Fundamental Fallacy** but does commence to show up in works I wrote after it. Even today, though, I find it difficult to understand why I should *have* to make it.

tered in any experience of anything smaller than approximately 2^{-128} times the diameter of the hydrogen atom);

(2) *Potentially* substantial substrata (*i.e.:* pure potency, which is the same as saying: that the internal characteristics of which are encountered anytime a human being is aware of empty space);

AND

(3) *Insubstantial* substrata (*i.e.:* matter, which is the same as saying: a line of tension generated in space by an actually substantial substratum whenever that actually substantial substratum acts in the presence of a drag inducing differential between its actuality and the potentiality of the potentially substantial substratum united to it).

In the above distinctions, we begin to see what is _the_ most dramatic illustration of the inadequateness and, hence, confused nature of the Aristotelian concepts of substrata in general and matter in particular: For Aristotle, the terms, "matter" (*i.e.:* matter *taken as a category*) and "that which exists potentially", mean the same thing. To rephrase it in the terminology of the Scholastics—a terminology based on Aristotle—the terms, "prime matter" and "pure potency", mean the same thing. In reality, though, they *do not* and *cannot possibly* mean the same thing. On the contrary, "pure potency" and "prime matter" refer to two kinds of substrata so radically different one from the other, that they do not even fall into the same *genus*.

"Pure potency" means the same as the term "space". It designates a "field" or "mantle" of potentiality which surrounds each and every *finite* unit of substantial substrata. To this mantle, we can attribute a diameter—a diameter which can be stated by comparing it to the diameter of the unit of substantial substrata which it is surrounding. For that reason, we can say that the diameter of the "mantle of potentiality"—relative to the diameter of the unit of substantial substrata it surrounds—varies inversely as the cube root of the fraction of potency currently converted by that unit into actuality. For example, let D_1 = the dia. of the mantle of pure potency, D_2 = the dia. of the unit of substantial substrata which it surrounds, P = the available potency, and E = the fraction of potency actualized. In that case, if E = 1/8 P, the cube root is ½, the inverse is 2/1, and D_1 = $2D_2$. Again, if E = 1/64 P, the cube root is ¼, the inverse is 4/1, and D_1 = $4D_2$.

Such, then, is the meaning of the terms "pure potency" and "that which exists potentially". "Matter", on the other hand, means no more than the line of force or tension which is generated at the center of that mantle as a result of the drag inducing differential between the mantle of space and the substantial substratum which that mantle is surrounding. As a result, whereas the mantle of space persists through each successive act of the substantial substratum it surrounds (changing only its *relative* diameter as it is either converted into the substantial substratum or reconverted into itself), the unit of matter (*i.e.:* the line of force) has being for only *one* act of the substantial substratum and is then, with the next act, annihilated and replaced by a new line of force.

Such, then, is the astounding difference between Aristotle's confused and inadequate concepts of "matter" and "pure potency" and concepts which have begun to achieve some appreciable degree of adequateness. Thus, in distinguishing between *substantial* and *insubstantial* substrata, there is born a kind of philosophy which transcends all prior philosophy more fully than the light of the sun transcends that of the moon. There is born a kind of philosophy which, on the one hand, gives us a *positively* valid concept of such terms as "pure potency", "that which exists potentially", "potentially substantial substratum", and "space", while, on the other hand, it gives us *adequate negatively* valid concepts of such terms as "substantial substrata" and "matter". For, on the one hand, it points the finger to what is commonly called "a vacuum" and says: "That vacuum there, which seems to be a sheer vacuity of all forms of being and reality, is a kind of reality having a kind of being (in the truest, primary and most definite sense of the word "being"), and it *is* what we signify by the terms 'pure potency' and 'that which *can be* a substantial substratum'." On the other hand, it gives to "substantial substrata" a definition which exactly fixes their location in the terms of the known diameter of the hydrogen atom, and gives to "matter" a definition which—when understood in all its fullness—will be found to be measurable and quantifiable by modern physics.

Up till now, when thinking of substrata, philosophers had to think *solely* in the terms of *substantial* substrata—substrata which are neither *present in* nor *predicable of* anything else.[5] The meaning of *in*substantial substrata, the relationship between substantial and insubstantial substrata, and how to think in the terms of

---

[5] **NOTE OF APRIL 25, 2002:** "Substrata" has already been defined as: "that of which all else is predicated, while it is itself not predicated of anything else". In short, "substratum" = "non-predicable".

an insubstantial substratum—these are realizations which have completely escaped every thinker prior to myself. Because I am the first individual in the history of fallen man to distinguish between substantial and insubstantial substrata; because I am the first individual in the history of fallen man to account for material objects in the terms of insubstantial substrata; because I am the first to produce a *positively* valid concept of "pure potency" by pointing the finger at space; because I am the first to produce an *adequate* negatively valid concept of matter (*i.e.:* matter as what the Scholastics would call "prime matter")—because of all these accomplishments, I am the first man in the history of fallen man to breach the wall between philosophy and science and to produce philosophical concepts capable of having merit before the court of empirical evidences. For those reasons, the system of thought given to the world through me is the first and only system of thought in the history of fallen man *truly* to deserve the title of "philosophy".

Such, then, is what I claim. If some must laugh at such manifest arrogance, then let them laugh. I am too old and too well blessed with physical and spiritual riches to care. Neither their jeers nor their indifference will have the slightest effect upon either my sleep or my bank balance or my relationship with God and friends. I have written what my conscience insists it is my duty to write, and that's all that *finally* counts.

(B)
STRUCTURE & FORMATTING:

What follows is not a single, coherent work. It is actually five different works. There is, nevertheless, a common thread running through each of those five. In each of them, my *foremost* objective was to draw a very clear and highly detailed picture of the differences between what I have called, on the one hand, *"substantial substrata"* and what I have called, on the other hand, *"insubstantial substrata"*.

Aristotle's Fundamental Fallacy is an essay I wrote in 1983. The other four works included herein are copies of letters which I wrote in response to the few pages of criticism which I was able to precipitate from the many universities and philosophical publications to which I sent copies of that essay.

I should perhaps point out that, except for the letter to Mr. Bigger of L. S. U., none of the letters, as given here, is a *duplicate* of the original. Each has been extensively re-written. Very little has been deleted or changed, but much has been added to each one. I make no excuse for that, since my main objective here is to explain the differences between substantial and insubstantial substrata—not to reproduce faithfully exactly what I said to the individuals in question.

I have, however, done the utmost to preserve the article, **Aristotle's Fundamental Fallacy**, exactly as it was in the original. Nevertheless, some changes have been made. They are all the results of the limitations of my 1983 computer and printer versus those of my current equipment. Basically, that means changes in formatting. As virtually all computer buffs know, the formatting abilities of computers and printers have changed dramatically over the course of the last 20 years. Other than changes in formatting, everything is exactly as it was in the original—including my stupid mistake of repeatedly confusing the plural, "substrata", with the singular, "substratum".[6]

Whenever, in the body of the four letters, I make reference to a page in **Aristotle's Fundamental Fallacy**, I will give the current page number in this book. That way, you the reader will be able to turn immediately to the page indicated and see for yourself what is in fact written there. It is possible that I have goofed here and there, and, in that case, the numbers given will refer to the original article, and you won't be able to find the indicated passage. I trust the reader will react with charity.

(C)
"HYPOSTASIS" VS "HUPOSTASIS":

Understandably, the computer and printer I used 20 years ago would not write Greek characters. That, combined with a poor choice of words on my part, created what may prove to be an aggravating problem.

If the word "hypostasis" is written in the *Greek* characters, it means "under standing" and indicates that which underlies any given individual such as this man or this horse.[7] Unfortunately, I made the bum choice of us-

---

[6] **NOTE OF MAY 3, 2002:** In Illustration #1, the equal sign and everything to the right of it is new. To each Aristotle quote's pg. #, I've added the pg. location.

[7] **NOTE OF APRIL 6, 2002:** Since first writing the above approximately 19 years ago, it has been brought to my attention that Aristotle and others also use the Greek term *"hupokeimenon"*, and it, too, implies what underlies something. Pardon me if decline to write it using Greek letters despite the fact that my current equipment would allow me to do so.

57

ing the *English* word "hypostasis" to mean "that which is undivided in itself and divided from all others"—a meaning the word took on down through the centuries following Aristotle. Taken in that latter sense, "hypostasis" is equivalent to such terms as "object", "compact mass", and "concrete individual". How, then, will the reader be able to distinguish between when I'm talking about "hypostasis in English" and when I'm talking about "hypostasis in Greek"? To remedy that difficulty, I have chosen to spell the word two different ways. When speaking of "hypostasis in Greek" (*i.e.:* "that which underlies"), I will spell the word with a "u", viz. "h*u*postasis". When speaking of "hypostasis in English" (*i.e.:* "concrete individual"), I will spell the word as it is normally spelled, viz. "h*y*postasis". The use of "h*y*postasis" is found in ***Aristotle's Fundamental Fallacy*** and in the letter to Mr. Scaltsas. The use of "h*u*postasis" is found in the first two letters.

(D)
MY OFFENSIVE STYLE:

I cannot deny that, at times, I wax quite hostile in my letters. Among educated people, it is quite fashionable to insist that *any* display of anger is reprehensible; consequently, most writers, will avoid all emotional display in their writings. I, though, couldn't care less what "educated people" say. Since our Lord Jesus Christ Himself did not avoid anger, I am not the least bit ashamed of that human emotion and do not hesitate to display it wherever I feel the other party has given me more than just cause to use it. Furthermore, I find that a bit of anger in the gut stirs the mind's ability to think and to express itself more dramatically.

At the same time, one should not read into my vehement passages, emotions which are not there. [8] My anger toward an opponent does not exclude gratitude toward that individual. Even in the midst of the fiercest wrath, I am still well aware of how much I owe to the four gentlemen who took the time to write to me. Had they wanted to do so, they could have refused to say anything to me whatsoever. Had they in fact done that, my mind would have had no reason to work out the clarifications which I did work out in these letters. In short, if my replies to the gentlemen in question have any value, much of that value is due to the men who gave me cause to write to them, and justice demands that I always acknowledge the debt which I owe to them.

But, do these letters in fact have any value? Will many people ever read what I have written here, let alone profit by it? I don't know for sure one way or the other. All I can say is that, knowing how badly I fare in all my attempts to express myself, *most* probably, very few people will ever bother even so much as to look at this book, and none will get a *great* deal out of what they read. To the few, though, who *will* attempt to read what follows (I am assuming you will all be friends reading out of regard for our friendship.), I can only apologize for my very difficult style of writing. I admit it

---

[8] **NOTE OF APRIL 28, 2002:** Would you believe it? In the vast majority of cases, where my words seem written in a raging fury, they were actually written while I was laughing my head off. Dear God! What fun it is to wade into the enemy with peals of thunder and lightning bolts! It's such fun, how is it possible *not* to laugh till your sides split? Try, then, to see my "vehemence" as a reason to smile rather than to frown.

SUBSTRATA

is obscure to say the least. My hope is, though, that your personal regard for me will somehow enable you to cope with the formidable obstacles which my lack of formal education is about to throw in your path, and, in doing that, hopefully, that charity toward me will empower you to draw from what follows at least some small tidbit of useful knowledge.

𝔓art #1:

LETTER TO
MR. ROBERT E. BUTTS

March 15, 1983

To: Mr. Robert E. Butts
Editor-In-Chief, "Philosophy Of Science"
Talbot College, Dept. Of Philosophy
University Of Western Ontario
London, Canada N6A 3K7

(A)
"SUBSTRATUM IS SUBSTANCE" VS.
"SUBSTANCE IS SUBSTRATUM":

Dear Mr. Butts:

I received your letter of rejection dated March 1, 1983. Included with your letter was a copy of the "referee's" commentary on my paper. You are the first and, to date, *only* one to afford me a critique upon my dissertation. For that, I am deeply indebted and cannot adequately express to you my gratitude for your generosity.

I must say, however, that I am exceedingly astonished by the extent to which the "referee" has misinterpreted and misrepresented my position. Your referee twice accuses me of saying: "Aristotle held the view that substance is substratum." I have read and re-read my paper in an attempt to discover where I might have inadvertently said that Aristotle held substance is substratum. Because your referee also accused me of saying that, for Aristotle, substance is *homogeneous* substratum, I also searched to see where I might have said that substance is *homogenous* substratum, and, thus far, have been utterly unable to discover where I said such things.

Most certainly, I repeatedly said that Aristotle held that substratum is substance and that matter is substance. **_But_**, in at least two places (See the quotes from Aristotle on page 250 of my article and the paragraph immediately preceding them.), I qualified that and explained that, for Aristotle, substratum is *potential* substance, and matter is *potential* substance.

Does your referee intend to deny that, for Aristotle, matter is *potential* substance???!!! In that case, then, how does he deal with such passages as these:

> It follows, then, that 'substance' has two senses, (A) the ultimate substratum, which is no longer predicated of anything else, and (B) that which, being a 'this', is also separable, and of this nature is the shape or form of each thing.
> ——*Metaphysics*; Book V; Chap. 8; 1017b: 24. **Great Books**; Vol. 8: pg. 538.

Since the substance which exists as underlying and as matter is generally recognized, and this is that which exists potentially, it remains for us to say what is the substance, in the sense of *actuality*, of sensible things.

——*Metaphysics*; Book VIII; Chap. 2; 1042b: 9. **Great Books**; Vol. 8: pg. 561.

If, therefore, your referee's complaint is that I'm saying <u>all</u> substance is substratum in Aristotle, then let me assure you of this: I deliberately refrained from what is generally called a "simple conversion" of the proposition. In other words, while saying that substratum is substance, I diligently sought to refrain from saying that substance is substratum. After all, I was well aware that, were I to affirm *both* the original proposition (*i.e.:* "All substratum is substance.") <u>and</u> its simple conversion (*i.e.:* "All substance is substratum."), I would be declaring "substance", "substratum", and "matter" are equivalent terms in Aristotle. That, I had no intention of doing, since I was well aware that, for Aristotle, "Substance"—when taken to mean *individual* substances—always implies substratum *plus form*. As I said on page 285, ". . . for him, whatever is being an *individual* is necessarily being a substantial substratum united with a form." Again, on page 247 of my article, I said: ". . . for Aristotle, 'substance' means far more than one mind can usually tolerate." I then went on to mention the multitude of meanings which "substance" has for Aristotle, and, on page 249 wrote:

"We must also bear in mind that, in addition to the formula, the concrete thing, the matter, the form, and the composite,

63

there is yet the essence and the universal, and they, too, are 'substance'.

Finally, on page 250, I *specifically* said:

"Since for the purposes of this dissertation, substance as substratum is our only concern, let us now **IGNORE ALL THE OTHER MEANINGS** Aristotle attaches to 'substance', and let us concern ourselves with what he says about substrata."

My dear sir, if I *specifically* said this paper was *not concerned* with all the *other* meanings of "substance" in Aristotle and was concerned *only* with "*substance as substratum*" (a phrase taken directly from Aristotle), how, in God's name, I beg you, could *any sane* man *possibly* accuse me of saying Aristotle held the view that substance is no more than substrata???!!! Did not the opening sentence of my paper say: "At issue here is one's basic ideas regarding *substance as substrata.*"???!!!

But, maybe I'm misreading your referee. Maybe he's charging me with not knowing that "substance" (*i.e.:* "*ousia*" in Greek) <u>never</u> to <u>any</u> extent implies "substratum" in Aristotle? If so, turn to chapter 4 of **Categories**, 1b: 25, where Aristotle says simple expressions signify substance, quantity, quality, relation, place, time position, state, action, or affection. Which of these does "substratum" signify if not *substance*?[9]

Since, however, I find it even more difficult to believe your referee's saying "substance" *never* means "substratum" in Aristotle, I *must* believe he's accusing me of saying *all* substance is substratum in Aristotle. In that

---

[9] **MAY 4, 2002:** On pg. 27, Owens quote says Aristotle's matter "is a principle in the category of substance".

case, what it boils down to is this: As I wrote, I was confident that, as long as I said only that all elephants are animals, none would be so mad as to conclude that I was also saying that all animals are elephants. Apparently, my confidence was foolish, for that is precisely the inference which your "referee" *brazenly* drew. Thus, he opens his "Referee's Report" saying:

> In the first ten pages of this paper it is contended that Aristotle held the view that substance is substratum, and that substratum is matter, and thus also that matter is substance. (c.f. page 251, para 3: "Matter is substantial substrata" (sic.)).

To begin with, sir, in the paragraph immediately preceding the quoted line from page 251, I stated:

> . . . let us define "substantial substratum" as: "any substratum which is neither predicable of another substratum nor incapable of existence apart from a subject". *With that definition in hand*, we can now state the Aristotelian position on substrata in the following three propositions: . . . . (2) Matter is substantial substrata.

Thus, I was not, at that point, even saying that matter is *substance*. I was only saying that, for Aristotle, matter is that kind of *substratum* which is neither predicable of another substratum nor incapable of existence apart from a subject. But, even if I *had* said that, for Aristotle, matter is *substance*, how does it follow that my statement is proof I am saying that, in Aristotle, substance is substratum? Thus, your referee, in an unbelievably tortured attempt to force from my writing the assertion that Aristotle considered "substance", "substrata", and "matter" equivalent terms, boldly summons forth my statement, "Matter is substantial substrata", as supposed proof that I assert: "Substance is substratum."

Good grief, sir! Must we go back and review the basic principles of logic which even a freshman is supposed to know backwards and forwards? Can it be your referee is really unaware that, in Aristotelian logic, you cannot automatically infer the simple conversion of what is generally called an "A" type proposition???!!!

I assure you, sir, that I nowhere asserted that, for Aristotle, "substance", "substratum", and "matter" are equivalent terms. In fact, section "G" of my article (pgs. 281-291) was a somewhat detailed attempt to show that one of the gratuitous assertions constituting the Aristotelian starting point was his assertion that there can be no *substance* unless *substratum* is conjoined with a form (pg. 285). I then went on to explain in some detail that one of the main differences between myself and Aristotle was *my* assertion that—*in direct contradiction* to *Aristotle's* assertion that *all* substance is *necessarily* a substratum conjoined with a form—that *some* substance (namely what I called "uncompounded hypostases") is a conjunction of substratum directly with being and without the intervention of a form (pg. 287). But, if I so frequently and so specifically stated that only I *myself* held that substratum is sometimes substance without the intervention of a form, how does one extract from my writing the assertion that, for Aristotle, substance means no more than substratum???!!!

Since, as it appears, it is necessary to remind your referee of some of the most fundamental principles of philosophy, let me digress here to point something out to you. "Substratum is substance" is a proposition composed of a subject and a predicate. "Substratum" is the subject, and "is substance" is the predicate. That means "is substance" is predicated of "substratum".

Now then, for Aristotle, it _is_ permissible to predicate the genus of the species, but it is _not_ permissible to predicate the species of the genus. That means it is alright to say "All elephants are animals," but wrong to say "All animals are elephants." As Aristotle writes:

> . . . for the species is to the genus as subject is to predicate, since the genus is predicated of the species, whereas the species cannot be predicated of the genus.
> ——**Categories**; Chap. 5; 2b: 19-20. **Great Books**; Vol. 8; pg. 6 lower right.

Therefore, to say "Substratum is substance" is to say that the word "substratum" indicates a _sub_-classification of the classification called "substance". In effect, it's the same as saying that the meaning of the word "substratum" is _one of_ the meanings included within the _several_ meanings of the word "substance". _Conversely_, to say, as your referee said, that "Substance is substratum" is to say that the word "substance" indicates a sub-classification of the classification called "substratum". In effect, your referee is saying that the meaning of the word "substance" is _one of_ the meanings included within the _several_ meanings of the word "substratum". _Clearly_, sir, your referee's statement, "Substance is substratum", is the _complete reversal_ of my statement, "Substratum is substance".

The question is this: In Aristotle, is it _in fact true_, that the meaning of "substratum" (_i.e.:_ "hupostasis" in the Greek) is _one of_ the meanings included within the _many_ meanings of "substance" (_i.e.:_ "ousia" in the Greek)? I say it _most certainly is_.

In what sense does Aristotle include *"hupostasis"* within *"ousia"*? I say that, for Aristotle, *both* *"hupostasis"* and *"ousia"* are each describable by *both* of these two critically important phrases:

(1)   that of which all else is predicable, while it is itself not predicable of anything else;

AND

(2)   that which is neither present in nor predicable of another.

If either you or your referee or anyone else wish to deny that Aristotle uses both of those phrases to describe *both "ousia" and "hupostasis"*, then you might as well try to deny that anything is anything.

To conclude this section, let me say this: Imagine a convention of philosophers. While addressing the group, a speaker says: "All elephants are animals." Immediately, an opponent leaps to his feet and calls out: "You heard what he just said. He said that all elephants are animals. Those very words out of his very own mouth convict him of trying to say that all animals are elephants. Who can listen to the man who proposes such manifest nonsense?" Imagine that the audience applauds his outcry. Faced with such "philosophers", what, I ask you, can the first speaker do but walk out the door and go elsewhere in search of men who are philosophers in something more than name only?

(B)
ARISTOTLE & THE 4 MEANINGS

OF "OUSIA":

Shortly after the last quote above from your referee, he writes:

> In particular, Aristotle's rehearsal of the popular notions of *"ousia"* (at 1028b 33 ff) is construed as a list of *his* interpretations of *"ousia"*: the fact that of the four three are subsequently rejected in the course of "Metaphysics VII" is ignored.

Come, come, sir! I am well aware Aristotle was merely rehearsing popular notions and that he immediately proceeded to reject some of them. But, that has nothing to do with my purpose in using the quote. Your referee asserts I made use of the quote as support for my supposed assertion that, for Aristotle, substance is substratum. But, look again at what I said on pg. 244 in the paragraph immediately preceding the quote from Aristotle. I said:

> Aristotle at least sometimes defines "substance" and "substratum" the same way. Of each, he says the difference between it and "everything else" is that it is not predicable of a subject (*i.e.:* stratum), while everything else *is* so predicable. The next two quotes confirm that.

As you can see, I *plainly* stated that the purpose of the quote was to show an instance of Aristotle referring either to substance or to substratum as *"that of which everything else is predicated, while it is itself not predicated of anything else"*. *That* was the phrase I told your referee to look for in the quote, and no sane person could *possibly* contend that I told anyone to look for the phrase "Substance is substratum" in the quote. Certain it is that, in the quote, Aristotle used the phrase at issue to refer to *substratum*, just as, in the other quote I gave, he used *the very same phrase* (*i.e.:* "that of which everything else is predicated, while it is itself not predicated of anything else") to refer to *substance*. As I (pg. 245) quoted from 1029a: 7:

> We have now outlined the nature of substance, showing that it is that which is not predicated of a stratum, but of which all else is predicated.
> ——**ARISTOTLE:** *Metaphysics*; Book VII; Chap. 3; 1029a: 7. **Great Books**, Vol. 8: pg. 551 upper right.

What more could I have done to make it clear which line in the quote was the important one and what it confirmed? For all of that, eventhough I had carefully warned your referee which line showed that, for Aristotle, "substratum" meant "that of which everything else is predicated, while it is itself not predicated of anything else", he apparently extracted the line, ". . . for both the essence and the universal and the genus are thought to be the substance of each thing, and fourthly the substratum", and then asserted I was using *that* line to support my supposed assertion that, for Aristotle, substance is substratum. What can one say of such incredibly blind misconstructions?

(C)

ARISTOTLE'S INCOHERENCE:

Your referee then goes on to say:

In addition, while most recent commentators agree that the apparent dissonance in "Metaphysics VII", between the conception of primary *"ousia"* as form + matter, the concrete individual, and the conception of it as form alone is due to the fact that in some contexts he is talking about definitions, which will look to form ("to ti an einai"), whereas elsewhere he is concentrating on reality (ontology if you will), which recognizes the concrete individual, this author slides the two conceptions together, happy to conclude that Aristotle's ideas are incoherent.

Where, I beg you, did I conclude that Aristotle's ideas are incoherent? What I said was this:

To extricate us from this impossible maze, Aristotle makes his famous distinction between the potential and the actual. Though the following two quotes may not answer all our questions about what "substance" means to Aristotle, they at least tell us that matter is *potentially* substance and substratum without *actually* being such, and that explains how the matter *is* the substance in one sense while it *is not* the substance in another sense. (pg. 249)

SUBSTRATA

Manifestly, I plainly stated that Aristotle's distinction between the potential and the actual *explains* how matter is substance in one sense and not substance in another. Surely, when a man says: "This *explains* it," he means there is no dissonance or incoherence. Does he not???!!!

I thought I clearly explained how the "apparent dissonance" is no dissonance at all. I thought I made it clear that, when Aristotle calls substance matter and form, he is speaking of the whole individual; when he calls substratum substance (I did not say: "When he calls substance substratum."), or calls matter substance, he is referring to the *potential* element in the whole; when he calls the form substance, he is referring to the *actual* element of the whole. In short, matter and form together are *individual* "substance", while matter and substratum are *potential* "substance" and form is *actual* "substance".

Though I made no reference to it in my paper, one must also bear in mind that, in Aristotle, there is a very important difference between what those of us who speak English might call: (1) "<u>a</u> primary substance", *versus* (2) "<u>the</u> primary substance". For Aristotle, in the **Categories**, "<u>a</u> primary substance" means the same as "individual substance", which is the same as saying "concrete individual". In the **Metaphysics**, however, "<u>the</u> primary substance" means "the substance of a thing", which, in turn, means "that which causes a thing to be what it is". Since, for Aristotle, the form (*i.e.: actual* substance) is always "that which causes a thing to be what it is", form is always, for Aristotle, "*the* substance of a thing" and "*the* primary substance". As such, it is, for Aristotle, "substance" in a pre-eminent way. (One might say that, for Aristotle, "substance", "<u>a</u> substance", and "<u>the</u> substance of a thing" are three very different variations on a central theme.)

72

If one understands all those distinctions between "individual substance", "potential substance", "actual substance", "<u>a</u> primary substance", and "<u>the</u> primary substance", then there is no harm in sometimes dropping the understood adjectives and, consequently, sometimes saying, perhaps to the horror of the *uninitiated*: "Matter and form together are substance, but matter is also substance and so is the form, but the form is more so substance than matter is." After all, we the *initiated* understand that, in one case we are speaking of the composite, in another case speaking of the basic parts, and in another case speaking of what is most responsible for making the composite what it is.

In many of his discussions of "substance", Aristotle merely did what, *for the sake of economy*, most of us do virtually every day: He merely dropped what would normally be important adjectives (in this case, "individual", "potential", "actual", "<u>a</u> primary", and "<u>the</u> primary") whenever he was reasonably sure that the reader would know, from the context, which adjective was required and intended. For example, how often do we hear a particular speaker, in the course of a speech, say: "Man is flesh and blood"? Later on, in the same speech, he says: "Above all, man is an immortal soul." After a few moments, he then says: "Man is the union of body and soul, and is not truly man without such a union." For all of that, we are not horrified by all that *seeming* self-contradictoriness. For, we know quite well what he means and make no complaint about confusion in his speech.

That's what I *thought* I made clear. For all of that, your "referee" is "happy" to conclude that I was "happy to conclude that Aristotle's ideas are incoherent." How strange it is!!!

In closing this section, let me re-emphasize a point. The "apparent dissonance" (your referee's term) in Aristotle's treatment of "substance" is *not*, as your referee suggests, the result of a *two*-fold application of the term; it is, rather, the result of a *three*-fold application of that term. Rather than merely speaking:

(1)   sometimes of definitions (*i.e.:* what we might call "logical substance"),

AND

(2)   sometimes of concrete individuals (*i.e.:* what we might call "ontological substance"),

Aristotle is speaking sometimes of definitions, sometimes of concrete individuals, and:

(3)   sometimes of causes (*i.e.:* what we might call "etiological substance").

As "logical substance", "*ousia*" is a "Category". It is the *first* category of the ten categories of things and, *like "hupostasis"*, it is described by Aristotle as *both*:

(1)   that of which everything else is predicated, while it is not itself predicated of anything else;

AND

(2)   that which is neither present in nor predicable of another (*i.e.:* where "present in" means "incapable of existence apart from another").

As ontological substance, *"ousia"* is "the concrete individual" (your referee's term) such as this man Plato and this horse Bucephalus. It is describable as the compound of matter and form.

As etiological substance, *"ousia"* is, to use Aristotle's terms, "the substance of a thing" and "the primary cause of its being" (cf. 1017b: 22 & 1041b: 28). It is describable as that which *causes* a thing to be what it is.

That is the *three*-fold distinction which must be kept in mind—not the *two*-fold distinction made by your referee. Keep that *three*-fold distinction in mind, and there will be no dissonance in Aristotle's treatment of *"ousia"* and *"hupostasis"*, and, concomitantly, there will be no difficulty in understanding *my* assertion that, for Aristotle, *"hupostasis"* is *"ousia"* in the sense that, like *logical* substance, it is neither predicated of a subject nor incapable of existence apart from that subject.

(D)
THE SENSE IN WHICH MATTER IS UNIFORM FOR ARISTOTLE:

i.
PRIMARY VS. SECONDARY MATTER:

Your referee then says:

There appears to be in this section of the paper considerable confusion about the role of Aristotle's notion of matter in his philosophy. Even at its most basic level this is not construed by Aristotle as *uni-form*, as this author suggests. Two salient features of his view which are not taken into account in this paper are: a. that the simplest bodies consist of matter in-formed by some pair of the properties hot, cold, dry and wet. (Fire is hot and dry, for example.) Compound bodies are composed of configurations of these, in varying degrees of complexity of mixture and structure.

My dear sir, hot and dry are in no way either substances or substrata. They are by no means "that which is neither present in, nor predicable of, a sub-ject". They are "that which is predicable of a stratum". Does your referee think me so stupid as to assert that, in Aristotle, no substance can have more than one in-ternal characteristic? When I said that, for Aristotle, matter is uniform, I meant uniform as "that which is not predicable of another". I meant that, for Aristotle, there could be no such thing as a unit of substratum which was simultaneously: (1) that which is neither present in another nor predicable of another and (2) that which, while it is *not predicable of* another, *is pre-sent in* another.

By your referee's own admission, he is speaking of matter as "the simplest bodies" having "informed properties". As I just quoted him saying, ". . . the sim-plest bodies consist of matter informed by some pair of . . . properties . . ." Necessarily, that means he is speak-ing of what matter is in the terms of the characteristics it receives from its form. But, I am speaking of what matter is in and of itself without the intervention of a form, and certain it is that, in Aristotle, though matter and substratum receive everything else from a form, there is one thing they do not receive from a form, namely: their definition as "that which is neither present in, nor predicable of, another". In short, your referee is *manifestly* speaking of what *post*-Aristotelian philosophers call "*secondary* matter", while I am speaking of what post-Aristotelian philosophers call "*prime* matter".

I must say I feel compelled to wonder if perhaps your referee might be unable to think in the terms of universals. He certainly seems unable to approach the concept of matter taken in its most general, abstract, and extensive application (***i.e.:*** matter as what is called *prime* matter). To speak of "the simplest bodies" and "matter informed by some pair of properties" is undeniably to speak of *particular* matter, which is to say matter as "this particular material object" (***i.e.:*** matter as what is called *secondary* matter). Naturally, matter, taken as "this material object", is either hot or cold or dry or wet. But, matter, taken as a general concept with the widest possible extension, is described by Aristotle using these three phrases:

(1)   pure potency, (I know of no instance where Aristotle himself used this term, but his phrase "that which exists potentially" has the same meaning.)

(2)   that of which everything else is predicated, while it is itself not predicated of anything else,

(3)   that which underlies.

Taken in that widest and most *abstract* and *general* of senses (***i.e.:*** taken as *prime* matter), matter is nei-

77

ther hot nor cold, nor dry or wet; it is that which *can be* such things. As for what it *is* (as opposed to *can be*), it is neither more nor less than the general concept described by the three phrases detailed above.

Having said that, we must hasten to add that Aristotle also uses still another phrase to describe the general concept "matter". He calls it: "that which is neither present in nor predicable of another", and, as he himself explains, the phrase "present in" means incapable of existence apart from that other. The grand result is that Aristotle gives us two extremely important phrases in definition of the general concept "matter". They are:

(1)   that of which everything else is predicated, while it is itself not predicated of anything else.

AND

(2)   that which is neither present in nor predicable of another.

For Aristotle, these two phrases refer to the same concept. They are equivalent. Because, for Aristotle, there is no difference in the application of those two phrases, he cannot possibly divide the classification "substratum" into two radically different sub-classifications. For him, then, taken as a *general concept*, "substratum" is *"sui generis"*, which is to say that, for him, _all_ substrata, whether it be called either matter or anything else, is neither more nor less than "that which is neither present in nor predicable of another". That's what it _**is**_ for Aristotle *as a universal concept*.

Suppose, though, we challenge Aristotle. Suppose we say that *all* substrata are indeed describable as "that of which everything else is predicated, while it is itself not predicated of anything else". At the same time, though, we say *not all* substrata are describable as "that which is neither present in nor predicable of another". In that case, the phrase, "that of which everything else is predicated, while it is itself not predicated of anything else", now becomes the description of an *extremely* general classification which sub-divides into two *sub*-classifications which, though they are not as general as their *master* classification, are still very general in application. They are describable as:

(1)   that which is **NEITHER PRESENT IN NOR PREDICABLE OF** another,

AND

(2)   that which **IS PRESENT IN** **BUT NOT** **PREDICABLE OF** that other.

If there in fact *are* two different species of "that of which everything else is predicated while it is not itself predicable of another", what, then, shall we say of matter (***i.e.:*** matter taken as a universal concept with the widest possible extension)? Aristotle will say that matter too is "that of which everything else is predicated while it is not itself predicable of another", and I will agree with him on that. He, however, will go on to say that matter *is* **_also_** "that which is *not* present in a subject", and he will add: "By being 'present in a subject' I do not mean present as parts are present in a whole, but being incapable of existence apart from the said subject." The minute he says that, Aristotle and I part company. For, I *admit* that matter *is* "that of which everything else is predicated, while it is not itself predicable of another", but I *deny* that matter (taken as a universal concept with the widest possible extension) is "that which is *not* present in a subject", and, in saying *that*, I *affirm* that matter <u>is</u> a kind of substratum which is *incapable* of existence apart from a more independent kind of substratum. Nevertheless, because matter *is* "that of which everything else is predicated while it is not itself predicable of another", it remains a kind of *substratum*, and that necessarily means it retains a kind of independence which, though it is not complete, is still a very pronounced independence when compared to accidents.

Now then, if matter (taken as a universal concept with the widest possible extension) *is not* "that which is neither present in nor predicable of another", then, how, in the Aristotelian system, does an individual substance (such as this man Plato or this horse Bucephalus) manage to be "that which is neither present in nor predicable of another"? In other words, if the matter of Plato's body is not included in the classification "that which is *neither* present in *nor* predicable of another", then how is *Plato* included in that classification *in the Aristotelian system*?

Is that a critical question *demanding* an answer? It most certainly is. After all, Aristotle told us:

> Substance, in the truest and primary and most definite sense of the word, is that which is neither predicable of a subject nor present in a subject; for instance, the individual man or horse.
> ——*Categories*; Chap. 5; 2a: 11-13.

If, therefore, we cannot show how, in the Aristotelian system, Plato is included in the group "that which is neither present in nor predicable of another", then we cannot show how, in the Aristotelian system, Plato manages to be an individual substance; we cannot show how Plato manages to be a completely independent individual; and the whole Aristotelian system falls to the ground.

Clearly, in his attempts to explain how each individual substance is what it is, Aristotle gives us only *two* principles: matter and form. Of these two, form is merely that which actualizes the potency of matter; consequently, form cannot cause an individual substance to be "that which is neither present in nor predicable of another" *save* by actualizing matter's capacity to be "that which is neither present in nor predicable of another" (I am here ignoring, *for economy's sake*, the *Christian* doctrine that the form of man is an immortal soul.). But, if matter *does not have* such a capacity, then it follows necessarily that form cannot actualize what is not there to be actualized. With both matter and form excluded as explanations, that means that, as long as Aristotle leaves us with nothing more than matter and form, then he leaves us with no way whatsoever to explain how a physical object could be "that which is neither present in nor predicable of another", and that means his doctrine of matter and form leaves us with no way to explain either how individual substances manage to be substances or how Plato, Bucephalus, you, or I manage to be completely independent individuals.

Very well, then! If matter (*taken as a universal concept with the widest possible extension*) is <u>not</u> that which is neither present in nor predicable of another, then how *does* it connect with that which <u>is</u> neither present in nor predicable of another. To put it another way, how does the <u>in</u>substantial substratum called "matter" connect with a *substantial substratum*?

That, of course, is what I tried to explain in my discussion of the three kinds of substrata. Apparently, not a single, solitary person, at the more than forty universities and philosophical publications to whom I sent my dissertation, was able to comprehend my explanation.

ii.
SIMPLE & COMPOUND SUBJECTS, SIMPLE & TRIUNE OBJECTS, & SUBSTANTIAL FUSION:

Let me, therefore, make here another attempt to explain what I said. To do that, let me explain the difference between: (1) a simple subject, (2) a compound subject, (3) a simple object, (4) a triune object, and (5) a conglomeration. Be sure to note that #1 is a simple _subject_ and #3 a simple _object_. For me, there's an important difference between "subject" and "object".

To start, let us assume that the term, "subject", is equivalent to each of the following terms:

(1)   that of which all else is predicated while it is itself not predicated of anything else;

(2)   a non-predicable;

(3)   a stratum;

(4)   a substratum.

A simple subject is one which cannot be destroyed. It cannot be destroyed, because it is not composed of any parts. In other words, no matter what you do to it, you cannot cause it to breakdown into two or more other subjects.

A compound subject is a subject composed of parts (*i.e.: potential* parts). It is composed of parts because, if you injure or destroy it, what was before *one* subject will suddenly yield *more* than *one* subject. For instance, the helium atom is composed of two hydrogen atoms. While in the helium atom, the two hydrogen atoms forfeit their individual identities and become two parts of the helium atom's identity. But, if fission occurs, the two hydrogen atoms will separate; the helium atom will cease to be; and each of the two hydrogen atoms will return to being an individual subject called a hydrogen atom. I realize this is not the most perfect example in the world, but it will suffice.

Before moving on, let us say one other thing about compound subjects. Wherever two or more subjects are so joined together as to cease being subjects and to become parts of the new subject formed by their union, that process shall be called "*substantial* fusion". "Substantial fusion", therefore, shall mean: "any union of two or more subjects which downgrades them from subjects into the parts of another subject". It is extremely important to remember that definition of *substantial fusion*.

A simple object is any subject, whether simple or compound, which is *not* disjoined within itself but *is* disjoined from all other subjects. To put it another way, a simple object is any subject having extremities which are in no way included in, or one with, the extremities of any other subject.

Individual material substances, such as the man Aristotle, appear to the human eye to be *simple* objects, but they are *not* such. They, I propose, are *triune* objects.

What is a *triune* object? A triune object is an *object* because it is without disjunction within itself while being disjoined from all others. It is *triune* because it is a fusion of three subjects which, rather than *losing* their individual identities to the object formed by their fusion, *retain* their individual identities. They, so to speak, become the "slaves" of the object formed by their fusion (and, thus, they forfeit their *liberty* to that object), but, like all *slaves*, they are by no means merely more of the internal characteristics of their slaveholder. Put technically, a triune object is that the extremities of which are either entirely, or at least mostly, common to three different subjects.

iii.
SIAMESE TRIPLETS
VS. CONGLOMERATES:

To explain what that means, let us fashion an imaginary monstrosity. Imagine a set of Siamese triplets. As we look at them, we see one perfectly normal body from the neck down. That is to say there is one torso sporting two arms and two legs in the usual places, and, if we probe the interior, there is one heart, one pair of lungs, one set of vital organs, one stomach, one colon, one urinary tract, and so forth. As for the head, there is a strangely shaped brain inside of it and three faces ringing it. In other words, as you walk around this apparition, you see first one face with its own two ears, two eyes, nose, and mouth, then a second face with *its* own two ears, two eyes, nose, and mouth, then a third face with *its* own two ears, two eyes, nose, and mouth, and, finally, the first face again.

We ask the first face: "Who are you?"

He replies vigorously: "I am Edward The Great."

We ask the second face: "Who are you?"

He replies timidly: "I am Norbert The Not-So-Great."

We ask the third face: "Who are you?"

A strangely hollow and lethargic voice moans slowly and with great difficulty: "I . . . am . . . Christopher."

What is in front of us? Is it *one* subject? By no means! It is three different persons; and yet, they are so totally united that we can't tell where one ends and the other commences. In other words, though they are three subjects, each subject's extremities coincide with the extremities of the other two, and thus the three subjects are truly only one object. That is to say, there is but one *body* before us, and it is without disjunction within itself while being disjoined from all the other bodies of all the other people around it. Still, that *mostly* normal, *one body* is *simultaneously* the body of three very different people.

We say to a famous surgeon: "Doctor, can't you perform an operation designed to remedy this insane situation? Use your knife to separate them."

He replies: "These three individuals are joined so extensively, that were I to use my scalpel, I *might* indeed separate them. But, Norbert and Christopher would *definitely die*, while Edward would be reduced to a vegetable for life. If any one of them is to live any kind of normal life, all three must be left to share the same one body."

To understand the above more fully, contrast it with a conglomeration. Imagine a swarm of bees hanging from a limb. To the naked eye, it looks like a single object, since it *appears* to be (*i.e.:* at first glance is) a single compact mass undivided within itself and divided from all other compact masses. The minute you pass your hand through it, though, each one of the thousands of bees forming the compact mass will readily and easily separate from the compact mass and go along its way just as healthy and capable as it was before.

In short, a conglomeration is only an *apparent* object. It is not a real object because each and every one of the subjects joined in that apparent object can readily leave and can do so with no loss to itself or to its fellows.

Here, then, is what constitutes a *triune* object. Like any object—whether real or apparent—it is a "compact mass" without disjunction within itself while being disjoined from all others. *Un*like a *simple object*— but *like* a conglomeration—the subjects joined together in the *triune object* retain their individual identities, despite the fact that they are so totally united and intimately bonded as to share virtually all the same, identical extremities. *Un*like the case with *either* a conglomeration *or* a *simple* object, if you attempt to break the *triune object* apart into the subjects in that object, all of those subjects will either cease to be or will be completely immobilized.

In short, a triune object is any object the extremities of which are shared by three different subjects none of which is either a part or a characteristic of any other subject but which, nevertheless, cannot be safely separated from the other two. Shorter still, a triune object is a sub-species of the species "*non*-homogeneous object" of the genus "object" (The other species is "homogeneous object" which is the same as saying "simple object".).

iv.
HYPOSTATIC FUSION:

Note another very important point: Wherever two or more subjects are so joined together as to form a single object in which they remain individual subjects incapable of being safely separated, that process shall be called: "*hypostatic* fusion" as opposed to the *substantial* fusion which downgrades a subject into a mere part. "*Hypostatic* fusion", therefore, means: "any union of two or more subjects which leaves each subject a subject (*i.e.:* a non-predicable) but renders each subject incapable of being safely separated from the other subjects in that union."

Go back, now, to our Siamese triplets. We ask Edward: "How do you manage in one body? What if you want to do one thing and one or both of the others want to do something else?"

He replies: "There's no problem there. I've got all the power. They do what I tell them to do, or else."

We say to Norbert: "Are you going to let that bully push you around?"

He replies: "I have to. I don't dare fight back. He's got all the power, so I just go wherever Edward goes and do whatever he does."

We say to Christopher: "What about you?"

A familiarly lethargic voice drones laboriously: "I . . . go . . . along . . . with . . . the . . . majority."

Edward is an instance of *actually* substantial substrata (*i.e.:* that which *actually is* that which is *neither* present in *nor* predicable of another). Norbert is an instance of *potentially* substantial substrata (*i.e.:* that which *can be* that which is *neither* present in *nor* predicable of another). Christopher is an example of *in*substantial substrata (*i.e.:* that which *is present in* substantial substrata but is *not predicable of* that substantial substrata). Together, they are *one object* but *three sub-*

jects. Each is a subject because, notwithstanding their inclusion in the same one set of extremities, no one of them is in any way predicable of the other two.

The first subject (***i.e.:*** Edward = actually substantial substrata) is *not present in* any other subject, eventhough it is *hypostatically fused to* the other two subjects. In saying the first subject is "not present in any other subject", I mean this: If you physically disjoin him from the other two as totally as he is disjoined from you and me, he will not immediately suffer annihilation.

The second subject (***i.e.:*** Norbert = potentially substantial substrata) *is present in* the first subject. In saying that, I mean this: If, without first changing him radically, you physically disjoin him from Edward, he will instantly suffer annihilation.

The third subject (***i.e.:*** Christopher = insubstantial substrata) *is present in between* the first and second subject, which is to say one of its two extremities is in Edward while its other extremity is in Norbert. Disjoin Christopher from either Edward or Norbert, and he will instantly suffer annihilation, and there are no circumstances whatsoever under which that outcome could ever be reversed.[10]

---

[10] **NOTE OF APRIL 22, 2002:** In "Haasian" cosmology, the first subject represents the actuality of either a form or a generator; the second represents the potentiality of either a form or a generator; and the third represents a line of force generated by a generator when its actuality acts in the presence of the drag inducing differential between its actuality and its potentiality. Taken in its strictest, truest, primary, most definite, and unqualified sense, "matter" signifies one or more of those lines of force. Matter is thus the most unusual and dependent of the *three* kinds of substrata. How different that is from Aristotle for whom all substrata are of the same *one* kind and, therefore, are uniform and homogeneous!

Finally, though there are three *subjects* in the one *object*, one of those subjects exercises undisputed sovereignty over the other two. Such, then, is *my* picture of "matter" (*i.e.: potential* primary substance taken as a universal concept).

v.
HYPOSTATIC CONJOINED WITH SUBSTANTIAL FUSION:

Consider again our Siamese triplets. Without our being able to see it, a "soul" comes along and *substantially* fuses with Christopher. Because it is a *substantial* fusion, Christopher ceases to be. A *new* subject is formed composed of Christopher (*i.e.:* insubstantial substrata) and the newly arrived soul. Note well what I said: The *substantial* fusion is between the soul and Christopher. It in no way affects the *hypostatic* fusion between Edward, Norbert, and what was previously called Christopher. As a result, we still have *three hypostatically* fused subjects (*viz.:* Edward, Norbert, and the soul which is now substantially fused to what used to be Christopher) being *one* object. It's just that one of those three *hypostatically* fused subjects is now a *compound* subject composed of two *substantially* fused subjects (*viz.:* Christopher and a soul) each of which is now downgraded to the status of parts of the whole formed by the substantial union of Christopher and the soul.[11]

---

[11] **NOTE OF APRIL 22, 2002:** These days, I subdivide substantial fusions into two kinds: essential and non-essential. What is the difference between them? In all substantial fusions, a subject ceases to be a non-predicable and becomes that which can be predicated as a part of another. In substantial fusions which are *essential*, a subject fuses with, and be-

SUBSTRATA

We talk again to the face we called Christopher, and we say: "How are things going, Christopher?"

An extremely powerful voice shoots back: "Christopher is no longer. I, Michael, am now all that Christopher once was, and I am also exceedingly more than that. For, Christopher was a wimp, but I am a mighty warrior, and I now rule this body."

Immediately, we say to Edward: "Edward, did you hear that? Somebody's challenging your power."

Edward replies with a shaky voice: "That's right, and I don't want to fight with Michael. I will obey his commands."

---

comes predicable of, another whose internal characteristics are basically the *same* as its own, and that *similarity* is due to the fact that each is the same kind of substratum. To put it another way, that substantial fusion is essential because the internal characteristics (*i.e.:* the ontological essences) of the fusing subjects are not readily distinguishable from one another; and so, in a sense, each forfeits its very essence itself to the essence of the whole. It's the natural result of fusing like with like. For example, when two or more lines of force fuse together to produce a visible object, that is an *essential* substantial fusion. In substantial fusions which are _non_-essential, a subject fuses with, and becomes predicable of, another whose internal characteristics are *radically different* from its own, and that *radical _dis_*similarity is due to the fact that each is a different kind of substratum. To put it another way, that substantial fusion is non-essential because the internal characteristics (*i.e.:* the ontological essences) of the fusing subjects remain readily distinguishable from one another; and so, neither—*to any extent whatsoever*—forfeits its very essence itself to the essence of the whole. It's the natural result of joining two things different even from the standpoint of the kind of substrata they are and the kind of being they possess. For example, when an immortal soul fuses with, and becomes the form of, a visible object, that is a non-essential substantial fusion.

vi.
MATTER + FORM VS. FORM + MATTER HYPOSTATICALLY FUSED TO SUBSTANTIAL SUBSTRATA:

What our little scenario is telling us is this: Aristotle is right in saying that each and every *individual primary* substance (such as this man Plato or this horse Bucephalus) *is* matter and form, because *only* matter and form are parts of the individual primary substance, which is to say each individual primary substance is, to use my terminology, a "compound subject" produced by the "*substantial* fusion" of matter with a form. **_But_**, Aristotle fails to understand the complex nature of the matter which, by *substantially* fusing with the form, produced the individual primary substance. He fails to understand that the matter—which *substantially* fused with the form—is also *still* fused *hypostatically* with what I call actually and potentially substantial substrata. In short, it is grossly insufficient to say that individual *material* substances are each *matter plus form*; it is absolutely necessary to add a critical phrase and to say individual *material* substances are each a form plus:

MATTER WHICH IS HYPOSTATICALLY UNITED TO SUBSTANTIAL SUBSTRATA.

Come back now to the original question: How does Plato, or any other individual substance, manage to be "that which is neither present in nor predicable of another"? The answer is that neither he nor any other individual *material* substance is such *in and of himself,*

91

because nothing which is rightly predicable *as a part of Plato's* **body** can be described as "that which is neither present in nor predicable of another". **_But_**, the matter which *is* a part of Plato's body itself *is hypostatically* fused with "that which is neither present in nor predicable of another". In short, Plato—like all individual primary substances *in the material world* (**_i.e.:_** Plato in so far as he is some particular, sensible mass of flesh, blood, bone, etc.)—manages to be "that which is neither present in nor predicable of another" *only by means of a subservient subject which is only hypostatically fused to his material side.*

In conclusion, then, **_if_**, as Aristotle says, "to be a substance" means "to be that which is neither present in nor predicable of another", then _no_ individual *material* substance (**_i.e.:_** no individual considered solely from the standpoint of his body) is a substance, because no individual *material* substance is "that which is neither present in nor predicable of another". At least, no individual's material side (**_i.e.:_** body) is such *in and of itself*; it is such only by means of a "Siamese" appendage—an appendage joined to it in a most unusual way. Oppositely, if a human being be considered from his _im_material side (**_i.e.:_** his soul), then each human being is—*in and of a part of himself!*—"that which is neither present in nor predicable of another".

In closing this section, let me say this: Aristotle understandably *could not* have, and *did not* have, even a *remotely* accurate idea of the complicated nature of either substrata in general or matter in particular. For him, they are simply "that which is neither present in nor predicable of another", and that's all there is to it. He made such a gross oversimplification because, in his day, there was virtually nothing whatsoever to suggest either the need for, or the content of, the idea of a triune object, which is to say a (to use common sense parlance) triune "piece of substratum" which, while it is *one object*, is *three subjects*, which is to say three kinds of substrata each *radically* different from the other two and in no way predicable of the other two but, nevertheless, inseparable from them. [12]

---

[12] **NOTE OF APRIL 6, 2002:** From the standpoint of his *soul* (*i.e.:* his *spiritual* side), Plato (as with every human being) <u>can</u> rightly be described as "that which is neither present in nor predicable of another". That's because the "stuff" of which his soul is "composed" (*i.e.:* the substratum which is being his soul) is a kind of substratum which can avoid annihilation without being intimately bonded to some more independent kind of substratum. Oppositely, from the standpoint of his body (*i.e.:* his *material* side), Plato (as with every human being) can <u>not</u> rightly be described as "that which is neither present in nor predicable of another". That's because the "stuff" of which his body is composed (*i.e.:* the substratum which is being his matter) is a kind of substratum which can **not** avoid annihilation without being intimately bonded to some more independent kind of substratum. Now then, if "existence" means the kind of *being* characteristic of that which can avoid annihilation even without intimate bonding to another, then, since Plato's material side (*i.e.:* the matter composing Plato's body) cannot avoid annihilation without intimate bonding to another, Plato's material side **DOES NOT ITSELF HAVE EXISTENCE**. On the contrary, the only reason why Plato's material side is not intimately bonded to all the other material bodies around him is that the matter of his material side is intimately bonded to numerous instances of a far more independent kind of substratum—instances we can rightly call the ultimate *generators* of the material and sensible world. Each of those *generators* is <u>so</u> independent it can *never* be *predicated* of Plato (whether of his body or his soul) and can never be made to become intimately bonded to another one of its fellow generators. Though enclosed within the extremities of Plato's body and intimately bonded to the

(E)
NO CORPSE IS A HUMAN BODY:

Your referee writes:

> A corpse is only inexactly named as a
> 'human body'; strictly it is no longer such.
> To contend then, that body is substance,
> and that substance is homogeneous sub-
> stratum is totally to depart from the Aris-
> totelian conception of the universe.

---

pieces of matter comprising Plato's body, the generators are
no more parts or internal characteristics of Plato's body than
are the parasites swimming in his blood stream. That necessarily means that the *existence* of Plato's body (*i.e.:* the existence of each and every piece of the matter composing his
body) is *outside* of Plato's body. To put it another way, the
capacity of *his* body to escape either annihilation or cohesion
to someone *else's* body is *extrinsic* to his body's matter. From
the standpoint of what is *truly* his body's matter, the capacity
to avoid annihilation and coalescence is an *external* rather
than an *internal* characteristic. This notion that the existence
of material things is outside of them is perhaps the one Pla-
tonic notion most vigorously rejected (and ridiculed) by Aristotle. Unfortunately for Aristotle, if the "Haasian" system
proves to be basically correct, then what, in Plato, Aristotle
denounced most radically, becomes, instead, one of the most
prescient leaps forward in the intellectual history of humanity. Indeed, like those few Greeks who said the Earth rotates
around the Sun, Plato will suddenly be seen as a man about
2,000 years ahead of his contemporaries—Aristotle included.

With that, I agree wholeheartedly. The problem is that your referee seems to think that I assert Aristotle held that the body of man is man by virtue of nothing more than the fact of it's being a homogeneous substratum, and I am at a loss to discover how he read such a twisted interpretation into what I wrote. If I maintained what he *alleges* I maintain, how could I even begin to distinguish between a dead body and a living one? Except perhaps for six years of atheism, I have been a Roman Catholic all my life, and I am supposed to be ignorant of the fact that death means the immortal soul, as form of the body, has departed???!!!

Without any doubt, for Aristotle, the dead body is no longer the *same* substance it was when it was alive; nevertheless, it is still *some* substance or a conglomeration of substances, and, whether we say the dead body is still some *one* new substance or *many* new substances, for Aristotle, each substance attributed to the dead body is *actually* a form imposing certain qualities, quantities, and so forth upon a substratum—a substratum defined as "that which is neither present in nor predicable of another". The form is, for Aristotle, the means by which the "concrete individual" has a particular series of characteristics such as quality, quantity, and so forth. *But*, the *substratum* is most certainly, for Aristotle, the means by which the "concrete individual" is neither present in, nor predicable of, another subject.

I *never* said that, for Aristotle, the body of a man manages to be "a man" by virtue of the fact that it is an homogeneous substratum. I was never so mad, stupid, blind, and misinformed as to make so manifestly ridiculous an assertion. What I said was that, for Aristotle, the body of a man—whether it is still a living man or is now a dead body, or a hunk of meat, or a zombie, or merely dog food or whatever else—is included in the general classification "**THAT WHICH IS NEITHER PRESENT IN NOR PREDICABLE OF ANOTHER**" by virtue of the fact that it is, *for Aristotle*, solely, wholly and entirely *underlain* by that *one* kind of substratum which *Aristotle defines* as "**THAT WHICH IS NEITHER PRESENT IN NOR PREDICABLE OF ANOTHER**". In short, I was *always* referring *solely* to what makes *any* particular material object a member of the general classification called: "**THAT WHICH IS NEITHER PRESENT IN NOR PREDICABLE OF ANOTHER**". I was *never* referring to what makes a particular material object *the particular primary substance or concrete individual which it is*. How can I possibly make it any clearer?

(F)
EXPLAIN THINGS IN THE TERMS OF THEIR ULTIMATE PARTICLES:

Your referee speaks of me as:

. . . . assuming with modern physics that the explanation of things is best approached by analysis of the simplest components of things.

My dear sir, that is utterly contrary to what I clearly said. I suggest your referee re-read page 290. I grant that *modern physics* seeks the explanation of things solely in the analysis of the simplest components of things, and, as I pointed out on page 290, Aristotle's theory of matter and form, because it attributes every internal characteristic of every material object to the form (except, of course, the definition of the matter and the whole as "that which is neither present in nor predicable of another"), leaves us with the opposite conclusion, namely: that we should seek the explanation of things solely in the analysis of forms. In short, the one extreme (*i.e.:* modern science) insists on analyzing nothing but the simplest components of things, while the other extreme (*i.e.:* Aristotle) insists on analyzing nothing but the forms. Deliberately steering a middle course between those two extremes, I said that, in my view:

> . . . it follows that mankind's struggle to know the world around it must be a three pronged investigation. On the one hand, we must attempt to discover what uncompounded essence is imparted to substrata by its inclusion in substantial activity, and that means we must search for the ultimate, indivisible particle of matter. On the other hand, we must strive to learn the mathematical formula essential to each of the forms. (pgs. 290 & 291)

Simply put, I clearly said that the explanation of things is best approached by concurrent analyses of *both* the simplest components of things *and* of the forms of things.

(G)
WHY DISTINGUISH BETWEEN SUBSTANTIAL & INSUBSTANTIAL SUBSTRATA:

Toward the end of his report, your referee wrote:

SUBSTRATA

It remains unclear to me what the advantages are to the view he offers, what explanations we might better be able to furnish, or what confusions we might more readily avoid. He suggests that the questions about insubstantial substrata are the most important confronting philosophy; he does not tell us why he believes this to be the case.

Admittedly, I did not bother to say why I believe such to be the case. Why should I have done so? Your referee himself denounces my paper as "too long". Why does he then immediately turn around and insist I should have made it even longer? Very well! I will, with your kind permission, now bother to state why "he believes this to be the case."

Aristotle's view of substrata is the same as the view of the man on the street. It is the common sense view. Substrata and matter are "that which is *neither* present in another *nor* predicable of another". That particular definition of matter and substrata is precisely what has destroyed epistemology and plunged philosophy into the hopeless quagmire of universal skepticism. I will explain why I say that.

As I direct my eyes at this sheet of paper, a certain visual image is impressed upon my retina and—in some way which still defies explanation by the scientific world—that visual image winds up as a visual sense image immediately given to a conscious mind. As my conscious mind gazes upon that visual sense image, I see certain colors having certain shapes. As I reflect on those shaped colors, I am pressed to ask myself: "Am I seeing what is *really* there to be seen, or is it all an illusion?"

Soon enough, my mind says to me: "There is only one way for this sense image to be *really* there: It *must* be that, in seeing this visual sense image, I am face to face with what some kind of stuff is being. That is to say, it *must* be that the internal characteristics I sense match the internal characteristics of that which *really* has being, and by '*really* has being' I mean that which has being *in the strictest, most proper, and unqualified sense* of the word 'being'."

For the sake of economy, the phrase, "that which has being in the strictest, most proper and unqualified sense of the word 'being'", shall be said to be equivalent to each of these terms: "stratum", "substratum", "subject", "stuff", and "that which underlies". That will now allow me to say this: Whatever has being in *any* sense of that term is either:

(1)   a substratum,

OR

(2)   the internal characteristic of a substratum,

OR

(3)   a logical classification devised by the human mind and supposed to apply to some one or more of whatever has being *in any sense of that term.*

The phrase, "internal characteristic of a substratum", shall mean the same as the term "accident". "Accident", therefore, indicates such things as the color, shape, texture, odor, flavor, weight, and size of the subjects which I experience in my sense images. It indicates such because, in the images given to my senses, those *are* the "internal characteristics of a substratum".

Can the accidents have being apart from their substrata? No, they cannot! For, if I say they have being apart from their substrata, then I say they have being apart from that which alone has being in the strictest, most proper, and unqualified sense of the word "being", and that amounts to saying that, apart from their substrata, they do not have being in the unqualified sense. Thus, by the very definition of *unqualified* being, accidents cannot *really* have being apart from their substrata. Therefore, we say this: The phrase, "internal characteristic of a substratum", and the term, "accident", are equivalent to each other *and* equivalent to the phrase, "that which is present in a subject". By "present in a subject", I mean incapable of having being—*in the strictest, most proper and unqualified sense*—apart from that subject.[13]

The phrase, "internal characteristic of a substratum" shall also mean the same as the phrase, "what a substratum is being". That is true because, for example, the accident, white, has being only because some subject *is white*, which is to say *is being white*. Therefore, white—as is *every* accident—is "what a substratum is being".

The phrase, "a logical classification devised by the human mind etc.", shall be equivalent to the phrase, "that which is predicable of another". For the sake of even greater brevity, it is equivalent to the term "a predicable".

So then, the phrase, "whatever has being in *any* sense of that term", has *three* different applications,

---

[13] **NOTE OF APRIL 7, 2002:** Remember that "apart from that subject" means apart from *intimate bonding* (**i.e.:** cohesion) to that subject.

namely: substrata, accidents, and predicables. "Substrata" means the same as either:

(1)   thing in the strictest, most proper, and unqualified sense of the word "thing",

OR

(2)   that which has being in the strictest, most proper, and unqualified sense of the word "being".

The terms, "accidents" and "predicables", however, indicate only *sub*-classifications of what can be described these two following ways:

(1)   thing in the broad and qualified sense of the word "thing",

AND

(2)   that which has being in the broad and qualified sense of the word "being".

To review, then, we have "that which has being in the <u>un</u>qualified sense" and we have "that which has being in the *qualified* sense". Only substrata have being in the <u>un</u>qualified sense, and, of the things which have being in the *qualified* sense, some are accidents present in a substratum, and others are logical classifications predicable of either strata, or accidents, or other predicables, or all three.

It must be admitted, though, that to use the term "predicable" to indicate logical classifications in the mind, is somewhat confusing, since it certainly seems that accidents and parts are also predicable of their substrata, as when we say that man is body and soul or that some men are white. In that case, one must divide "predicables" into "predicables in the unqualified sense" and "predicables in the qualified sense". Once we say that, it follows that whatever is a thing in *any* sense of the word "thing" (and whatever has being in *any* sense of the word "being") is either a non-predicable (*i.e.:* a substratum) or a predicable in either the qualified or the unqualified sense.

We have now laid the groundwork for the desired explanation. Let us, therefore, ask a question. What _is_ the strictest, most proper, and unqualified sense of the word "being"?

Aristotle appears on the scene and says: "Taken in the strictest, most proper, and unqualified sense, 'to be' means 'to be that which is neither present in nor predicable of another'." That certainly seems to accord with the list of three kinds of things which we have just drawn up, but it is actually an exceedingly disastrous one.

That Aristotelian definition of _un_qualified "being", "substance", "substratum", "stuff", "that which underlies", or whatever else you want to call it—that definition, I say, has always caused the *vast, vast, vast* majority of all human beings, whether _un_educated, educated, or *highly* educated, to fashion virtually the same one identical picture. In an attempt to "picture" a kind of "stuff" which truly _is_ "neither present in nor predicable of another", they first imagined a big hole in the middle of nowhere. By itself, this big hole is pure nothingness, which is to say *it itself* is a sheer vacuity of all forms of being, substance, substrata, stuff, and so forth. For all of that, it has length, breadth, and depth, and real things can move around in it.

And what are these *real* things? They are "little globules" (to use Locke's terminology) and "adamantine lumps of stuff" (to use Newton's terminology). Each "little globule" is utterly solid, devoid of any empty spaces, wholly impermeable, three dimensional, and cannot be reduced any further in size.

With that picture in mind, it's easy to be sure the internal characteristics sensed match those of a stratum. I say to myself: "If I am seeing a rust-colored, solid, smooth, iron sphere, it's because, in the object outside of my body and in the sense image inside of my body, there's a lot of little iron globules each of which is red all over, and they squeeze themselves together and stick to one another so tightly into a ball-like shape, that they leave no empty space between themselves anywhere within the confines of that ball-like shape."

For the *vast vast vast* majority of mankind—including even the most educated of men and the keenest of philosophers and scientists—that infantile little picture served as *the* basis of certitude from the dawn of man's intellectual history up to the seventeenth century. Some might try to deny it, but it would take little effort to bury them beneath an avalanche of evidences.

Why did so many acute thinkers find it necessary to think in the terms of "little globules" the instant they concluded that "to be" means, in the unqualified sense, "to be that which is neither present in nor predicable of another"? Let us answer that question as clearly as possible.

To say of any given subject that it is neither present in nor predicable of another, is to say that what *ultimately* underlies it is its own self, which is to say that its own potential parts are what give it being. To put it another way, it does not share its extremities with some *other* subject or subjects acting as the means by which it has being. In common sense parlance, it means that

103

whatever is within the confines of a given object is either a part of that object or a contaminant which does not belong there.

Imagine, then, a highly intelligent man looking at his visual image of a rust-colored, solid, smooth, iron sphere resting on a marble table. As he looks at his visual image, he thinks about the extra-mental reality which that visual image is supposed to represent. If he is to be certain that he _is_ seeing what he is seeing, then it must be that he is seeing the internal characteristics of that which is neither present in nor predicable of another. Necessarily, that means that, excluding perhaps a few, insignificant contaminants, whatever is within the confines of his visual image must be a part (*i.e.: potential* part) or "piece" (in common sense parlance) of that image—just as whatever is within the confines of the extra-mental, iron sphere must also be a piece of that iron sphere.

Obviously, if *every* iota of that which is within the confines of the iron ball is virtually the same as the entire iron ball, there can be _no_ question, that, in seeing the visual image which I _do_ see, I see the internal characteristics of that which has being in the strictest, most proper, and unqualified sense, which is to say I see the internal characteristics of that which is neither present in nor predicable of another. Manifestly, if the iron sphere and my visual image of it are both conglomerations of solid, rust colored, little globules of iron stuck tightly together, then there can be no question that—no matter *how small* a piece I may examine—I will *always* see a solid, rust-colored, little globule of iron, and that will *absolutely confirm* my certitude that I see what _is_. Again, as long as the *internal characteristics* of *every iota* of what's within the confines of both my sense image and the extra-mental original mostly duplicate those of the whole, then my ability—to be certain that I am face to face with the internal characteristics of something real—is preserved. Because the notion of the little globules is the *only* concept which fulfills those conditions, it has ever been the *only* concept to which most men have always clung with great desperation.

In review, then, to say that "to be" means, in the unqualified sense, "to be that which is neither present in nor predicable of another", is to say:

(1) Except for contaminants, whatever is within the confines of any individual substance is a part of that individual substance.

(2) No matter how small, every part of any given individual substance (**ex. gr.:** this man Plato) must have internal characteristics which are mostly the same as the internal characteristics which the whole of that given individual substance presents to the observer.

Onto the scene, come the seventeenth century scientists with their microscopes. And what did educated men see when they put the rust-colored, smooth, solid, stationary, iron sphere under their microscopes? They saw, within the confines of the whole, little pieces whose internal characteristics bore virtually no resemblance whatsoever to those of the whole. Some semblance of "solidity" was the only similarity they found, and that solidity was confined to unbelievably tiny pockets within the confines of the whole. To make matters worse, when they focused on those tiny pockets of solidity, they, too, proved to be mostly empty space.

SUBSTRATA

Since impermeability was the only familiar characteristic left, it became the accepted sign of where "the real thing" was to be found, and since those tiny little pockets of "the real thing" obviously sported none of the other internal characteristics of man's sense images, it was concluded that, for the most part (except perhaps for impermeability), our senses present us with what is the internal characteristics of *absolutely nothing whatsoever*. "All that we see, hear, taste, touch, and smell," lamented the Age Of Enlightenment, "is *non*-being." If that is true, of course, then all knowledge is purely subjective, and truth is nothing more than whatever appears convenient.

Such, then, is how universal skepticism came to pass. Try to explain sensed shapes, colors, and so forth, in the terms of the internal characteristics of a kind of substratum which is itself neither present in nor predicable of another, and science will readily and most demonstrably show you the futility of your efforts, and the result will immediately be an unbridgeable gap between the world immediately impressed upon the conscious mind and the world outside of it.

Put bluntly, all the confusion brought upon philosophy by Descartes, Berkeley, Hume, Locke, Kant, and Leibnitz is traceable to the fact that, following Aristotle and the common sense view of matter, they could not think of matter in the terms of a substratum which—while it *is present in* another substratum—is *not predicable of* that other substratum. Trying to explain what we see, hear, taste, touch, and smell, in the terms of what *substantial* substrata are being (*i.e.:* in the terms of the internal characteristics of substrata *neither* present in *nor* predicable of another), Kant and company could not avoid being startled into total subjectivism when they discovered no similarity whatsoever be-

106

tween the internal characteristics sensed and those of "the little globules".[14]

---

[14] **NOTE OF APRIL 7, 2002:** These days, I have a second and perhaps more satisfying way to explain the impetus behind the great search for "the little globules". Let's start the explanation with a few definitions. First, let's define "being" as meaning "involvement in activity" and "to be" as meaning "to participate in activity" or, if you prefer, "to be involved in activity". There are three ways to participate in activity—namely: (1) to be involved in activity after the manner of an agent (*i.e.:* to exist); (2) to be involved in activity after the manner of a *durable* patient (*i.e.:* to *in*exist indefinitely); and (3) to be involved in activity after the manner of a *non*-durable patient (*i.e.:* to *in*exist briefly) . Involvement in activity after the manner of an _agent_ means to be that which is actually performing the action. Involvement in activity after the manner of a _patient_ means to be that which is merely "going along for the ride". *Durable* patients go along for the ride indefinitely and may themselves eventually become agents. *Non*-durable patients go along for the ride only briefly and then are annihilated. Next, let's define "island mass" to mean "that the extremities of which are spatially separated from all other extremities". If you prefer, say: "that whose set of extremities does not cohere to any other set of extremities". Next, let's define "conglomerate" to mean "a collection of island masses which, for some known or unknown reason, at least rather persistently maintain a rather fixed and close proximity to one another". If you prefer, say: "a collection of island masses which, for some known or unknown reason, move, overall and for some time, in a rather uniform way". Finally, let's define "irreducible mass" to mean "that which immediately appears to be an island mass and, no matter how greatly magnified, never turns out to be a conglomerate". If you prefer, say: "that which is absolutely not merely a collection of island masses".

    With those definitions out of the way, we can now say this: *Manifestly*, no *conglomerate* can *ever* be an agent, which is to say can never be involved in activity after the manner of an agent, which is to say can never be that which actually performs this or that particular action, which is to say can never *exist*. For example, when a band marches down the street, it is not that conglomerate *itself* which is doing the marching; it is, rather, each of the players actually performing the activity called "marching". But, each of them is a conglomerate whose island masses are called molecules; therefore, it is each of the molecules in each of the players which is actually doing the marching. But, each of those molecules is a conglomerate whose island masses are called atoms. But, each of those atoms is a conglomerate whose island masses are called leptons and quarks. Well, then, is each of those leptons and quarks an irreducible mass? The scientists are not sure. What, then, *is* the agent actually doing the marching? The only way we can answer the question is to find an island mass which continues to be an island mass no matter how many times it is magnified. In other words, if we are to find something which *exists* and which *actually does* the various actions we daily observe, then we **must** find what Locke calls "the little globules" and what Newton calls "adamantine lumps of stuff", and that realization—if only subliminally—is what *drives* the quest for those "ultimate building bricks".

But, is scientific probing into ever smaller things the *only* way to find the illusive irreducible mass? Consider this: When you think, is it *you yourself* who are thinking. If so, then it is you who are actually performing that activity, and that makes you an irreducible mass engaged in activity after the manner of an agent. Do you deny or doubt that you yourself are performing your mental activity? If you *do indeed* either deny or doubt, then it is *you yourself* who are performing that mental activity called denying or doubting, and that makes you an irreducible mass engaged in activity after the manner of an agent, which is to say that makes you something which *exists* in the strictest, truest, primary, most definite, and unqualified sense. Yes, Monsieur Descartes! If it is indeed you yourself who think, then you exist in the most perfect sense of all—namely: you are an irreducible mass engaged in activity after the manner of an agent. Something

What happens, though, if we say that "to be"—*taken in its strictest, most proper, and unqualified sense*—can mean **EITHER**:

(1)   to be that which is **NEITHER** present in **NOR** predicable of another,

OR

(2)   to be that which **IS PRESENT IN** another but is **NOT PREDICABLE OF** that other.

The first thing that follows it this: Instead of having only *one* meaning (as it does in Aristotle), the word *"being"* (*taken in its strictest, most proper, and unqualified sense*) has *three* distinct meanings (**viz.:** 1: actually substantial being; 2: potentially substantial being; and 3: insubstantial being). At the same time, instead of having only *one* meaning (as it does in Aristotle) the word *"thing"* (*taken in its strictest, truest, most proper, and unqualified sense*) has *three* distinct meanings (**viz.:** 1: actually substantial thing; 2: potentially substantial thing; and 3: insubstantial thing).

The second thing that follows is this: Because there are *three* proper and unqualified meanings to the words "being" and "thing" (instead of only one as in Aristotle), it is possible to have a real being (such as this man Plato, this horse Bucephalus, or this iron ball) which—while it is a thing in the strictest, truest, most proper, and unqualified sense, and has being in the

---

other than you may deny it or doubt it; but, you yourself most certainly cannot, unless, of course, you are the irreducible mass engaging after the manner of an agent in the acts of denial and doubt.

strictest, truest, most proper, and unqualified sense—must, nevertheless, be so present in another thing that, were it separated from that other thing (*i.e.:* did it cease to *cohere* to that other thing) , it would instantly be annihilated. One thus has a subject which cannot exist apart from a certain second subject (or group of subjects), and, because both it and its supporting subject (or subjects) are truly *subjects*, neither can be predicated of the other; and yet, the larger subject will utterly cease to be, unless the smaller subjects—*without* becoming *parts* of the larger subject—consent to be its willing slaves lending it *their* ability to be *not present in* another.

Once it is possible to have a subject resting in other subjects which are *subservient* to it *without* being *parts* of it, the third thing which follows is this: It is no longer true that *whatever* is within the confines of the iron ball is *a part of* that ball. For, if the iron ball is *in*-substantial substrata, then much of what lies within the confines of the iron ball is:

**ANOTHER SUBJECT OR SERIES OF SUBJECTS
HYPOSTATICALLY UNITED TO THE IRON BALL
AND SERVING AS
THAT WHICH SUSTAINS IT IN BEING**.

The fourth thing that follows is this: This business of going ever deeper into the microscopic world within the iron ball, in order to find its "ultimate constituents" is a waste of time. Such efforts do not touch the *parts* of the iron ball; they touch only what can be called its "slaves"—*subservient appendages* which are only *hypostatically* (rather than *substantially*) connected to the ultimate constituents of the iron ball.

The above means that the molecules in the iron ball are *not parts* of the iron ball; the atoms in the molecule are *not parts* of the molecule; the leptons and quarks in the atom are *not parts* of the atom; and the rishons (as some would call them) in the leptons and quarks are *not parts* of the leptons and quarks. The molecules are—to speak figuratively—the "posts" from which the iron ball's matter is hung; the atoms are merely the "posts" from which the molecule's matter is hung, and so forth.

I say that science will eventually find that, in every material object, there is strung, between the molecules, an immense but definite number of lines of force which form a very intricate web of force filling the supposedly empty spaces between those molecules. That gossamer web of lines of force is the "stuff" of which the visible material object is made, and, in the case of a rust-colored, solid, smooth, iron sphere, the web truly is rust-colored, smooth, spherical iron. That web is *not*, however, solid *in and of itself*. Its solidity is *not **its** solidity*. It borrows that solidity from its "slaves"—from the subservient appendages hypostatically united to each of the two ends of each one of its lines of force.

The paramount point, however, is this: The web which is woven by the molecules in the iron sphere can also be woven, in a miniaturized version, by the molecules in the brain, and that web can then be presented to the conscious mind which then experiences the internal characteristics of that web. In other words, whereas Aristotle says our senses receive *only* the form *stripped* of all sensible *matter*, I say the senses receive *both* the form *and* the matter *stripped of the supporting appendages* (Having it own "slaves", the brain needs not the iron ball's slaves.). Since the senses receive the mat-

111

ter itself and gaze upon that matter's internal characteristics, that is how it comes about that—in feeling the lines of force in the *intra*-mental web reproducing in miniature the *extra*-mental web called "an iron sphere"—I come face to face with an extremely accurate, intra-mental reproduction of what the extra-mental iron sphere is being in itself. That's the way it *must* be, because that's the only way I can be sure that, in seeing what I see, I see what has being in the strictest, most proper, and unqualified sense of the word "being".

To summarize, as long as the world agrees with Aristotle that *no* substance (such as this man Plato) is ever *present in another*, there's no way to explain the internal characteristics we *sense* in Plato save in the terms of the internal characteristics of Plato's sub-atomic particles. Since the latter are too different from the former to allow such an explanation, the former are *nothing*, and universal skepticism is inevitable. But, the instant we distinguish between *substantial* and *in*substantial substrata, what's sensed can be explained in the terms of a substratum whose internal characteristics match those sensed, and universal skepticism vanishes. Then can we understand why the trees falling in the forest still make a mighty crashing sound even when no man or animal is there to hear it. It's because the "stuff" of which sense images are made is the same as the "stuff" of which *everything* material is made: *in*substantial lines of force persistently generated by the ultimate *substantial* constituents of the universe. As for the smaller "particles" of which we have long *thought* material objects are made, they are merely the "weavers" slavishly spinning *between* themselves the "webs of lines of force" which alone *truly are* the larger objects our senses experience.

Patently, the above is what is called a "copy" theory—a theory which says the sense image is a "photocopy" of the extra-mental reality. So what? A theory is not wrong just because others have grown use to scorning all theories of its kind.

Some will say I'm abusing Aristotle—that I'm taking, in an ontological sense, a phrase he used only in a logical sense. I reply, in the first place, that neither I nor anyone else will probably ever *prove* whether or not Aristotle himself took it in an ontological sense. In the second place, I say that whether or not Aristotle took the phrase in an ontological sense is of no importance whatsoever. The only important point is this: Most assuredly, the vast majority of those who came after him *did* take it ontologically.

Here again, however, I am no doubt wasting my time, just as I have been doing now for over twenty years. Nevertheless, it is my firm conviction that, if the grace of God will but co-operate with my efforts, I shall never quit trying to communicate my insights to the world around me. As I go on struggling, I do not do so with any hope that I shall one day *succeed* in communicating my ideas to the world around me. On the contrary, I am wholly convinced that I shall *never* succeed in accomplishing such. Either I am more stupid, confused, insane, and blind than my mind can conceive, or the world around me is such. It matters not which may be the case. The result is the same: I am certain that I and the world around me will never succeed in understanding one another one iota.

Why, then will I go on? I will go on because, when I die, I will be able to stand before my own self and say: "My self, you have no reason to attack me with hatred, disgust and brutal vengeance. I conducted myself well in a most courageous and noble way. Eventhough I knew that—because I myself am more incompetent than words can express—there was no hope whatsoever that I could ever succeed in bringing light to the world around me, nevertheless, I never ceased trying—not even with my last dying breath." In short, I go on for my own sake—for the sake of preserving my own sense of dignity and for the sake of achieving a more productive and satisfying relationship inside of myself and with myself. In the final analysis, such I seek to accomplish with no regard whatsoever for whether or not I have any chance of achieving a more productive and satisfying relationship with the world outside and around me.

(H)
ARISTOTLE'S DEFINITION OF A DEFINITION:

In conclusion, sir, let me say this: Aristotle defines substance in two very different ways. As I have shown, he repeatedly defines it the same way as he defines substratum. He defines it as: "that which is neither present in nor predicable of a subject". On the other hand, he most definitely "defines" substance as "substratum and form", or, as you might prefer to have it, as "matter and form". Aristotle specifically refers to each proposition as a "definition" of substance. But, if you examine his *definition of a definition* in **Metaphysics**, Book VII: Chap. 12, you will find that a definition—*in the strictest, most proper, and unqualified sense of the word*—is the genus and the last differentia. Now, then, which of the two "definitions" of substance given us by

Aristotle gives us the genus and the last differentia? I dare say the answer to that is obvious.[15]

As I showed in my article, Aristotle breaks down the genus "thing", the genus "being", and the genus "essence", into two species each. On the one hand there is "thing in the *strict, un*qualified sense", "being in the *strict, un*qualified sense", and "essence in the *strict, un*qualified sense". On the other hand, there is "thing *not* in the *strict, un*qualified sense", "being *not* in the *strict, un*qualified sense", and "essence *not* in the *strict, un*qualified sense". Furthermore, for Aristotle, "thing in the *strict, un*qualified sense" is also "thing *neither* predicable of another *nor* present in another" (*i.e.:* thing-in-itself); whereas, "thing *not* in the *strict, un*qualified sense" is: "thing both present in another and predicable of another" (*i.e.:* thing-in-another). By the same token, "being in the *strict, un*qualified sense" is also: "being *neither* predicable of another *nor* present in another"

_____

[15] **NOTE OF APRIL 27, 2002:** In speaking of substance as "that which is neither present in nor predicable of another", Aristotle is *effectively* saying this: "I define 'substance' as 'non-predicable non-inhering thing'. The genus is 'thing' and the last differentia is 'non predicable non-inhering' and by 'non-inhering' I mean not so present in another as to be unable to escape annihilation unless intimately bonded to that other. 'Substance' thus contrasts with 'accident' which I define as 'predicable inhering thing'." He then turns around and defines "substratum" the very same way. However, in the case of *substratum*, he *ought* to have said: "I define 'substratum' as 'non-predicable' thing. 'Thing' is the *order* and 'non-predicable' is the *family*. That family then yields these two *genera*: (1) *non*-inhering non-predicable thing, and (2) *inhering* non-predicable thing. That second genus admits of no species; but the first genus admits of these two species: (1) *actually* non-inhering non-predicable thing, and (2) *potentially* non-inhering non-predicable thing."

(*i.e.: "ens in se"*); whereas, "being *not* in the *strict*, <u>un</u>-qualified sense is: "being in another" (*i.e.: "ens in al-tero"*).

Clearly, to call substance and substratum "that which is neither present in another nor predicable of another", is to say it is of the genus "thing" and of the species "thing in the *strict* and <u>un</u>qualified sense", as opposed to the species "thing *not* in the *strict* and <u>un</u>-qualified sense". Therefore, to define substance as "that which is neither present in, nor predicable of, another", is to state what is the *strict* and *unqualified definition* of substance; whereas, to define substance as "matter and form" (or "potency and actuality" or "substratum and form") is clearly to state what is a definition only in a *broad* and *qualified* sense, since it makes no reference to either genus or specific difference and gives only the composition of the individual substance. In short, then, *even in the Aristotelian system*, the proposition, "Substance is matter and form," *is not a definition*. At least it is not such *in the strictest, most proper, and unqualified Aristotelian sense of the word "definition"*. As I quoted from Aristotle only a few pages earlier in this very letter:

> Substance, in the truest and primary and most definite sense of the word, is that which is neither predicable of a subject nor present in a subject.
> ——*Categories*; Chap. 5; 2a: 11-13

Throughout my article, I was well aware of the two radically different definitions of substance put forth by Aristotle. For that reason, I was *always* well aware that, though "substance" and "substratum" (*i.e.: "ousia"* and *"hupostasis"*) are equivalent terms for Aristotle *from the standpoint of his strict, unqualified definition* of those terms, they are not equivalent for him *from the stand-point of his broad and qualified definition* of substance. That is why, though I said Aristotle held substratum was substance, I deliberately refrained from ever saying that he held substance was substratum. To be even more critically exact, I said that, for Aristotle, substratum was *substantial*—meaning, by that, not that substratum was *substance*, but rather that, like substance, it is neither present in, nor predicable of, another. I then tried to show that Aristotle's insistence upon giving both sub-stance and substratum the same *strict* and *un*qualified definition (as opposed to the *broad* and *qualified* one) was unwarranted, since, although Aristotle could not conceive of how one could have a substratum which is *present* in but *not predicable* of another, **we can**. That, of course, is why I dwelt solely upon Aristotle's *strict* and *un*qualified definition of substance and substratum.

Have I perhaps made it all clearer to you now, Mr. Butts? No doubt, I have not. Let me, therefore, make one final attempt to explain what I am saying.

As you are no doubt well aware, Aristotle distin-guishes between what he calls "primary substance" and "secondary substance". Primary substance is what we might call "ontological substance". It is the *real* thing. It is this man Plato, this horse Bucephalus, this woman Helen of Troy, and so forth. Secondary substance is what we might call "logical substance". It is the mental classi-fication assigned to "ontological substances" by the hu-man mind.

Of course, as I pointed out earlier, "primary sub-stance" can mean either "*a* primary substance" or "*the* primary substance" in the Aristotelian system. As "*the* primary substance", "primary substance" means "the substance of a thing", which is the same as saying "that which causes a thing to be what it is", which is the same as saying "form". Obviously, I am here using "primary substance" to mean "a primary substance", and I have no reason to use it to mean "the substance of a thing".

If I point to a particular human being and ask what it is, then, if I am to state what it *in fact is*, I must give the names of its most basic parts and say: "That is matter and form." On the other hand, suppose we are discussing *definitions* and how we the thinkers *classify* real things. In that case, if I ask what "substance" is, I am no longer asking what *Plato* is or what *Aristotle* is; consequently, my answer will not give the names of *their* parts. On the contrary, I must name the common, universal characteristic which, wherever it is found, causes its host to be included in the logical classification called "substance". In this case, "substance" is, in the Aristotelian system, *most undeniably*:

"THAT WHICH IS NEITHER PRESENT IN NOR PREDICABLE OF ANOTHER".

Consider an example of what the above chain of thought implies. We point to an individual human being, and we say: "What is that?" If we are sharp, we reply: "That is a body united to an immortal soul." Next, we turn away from that individual human being and say: "What is man?" Again, if we are sharp, we say: "'Man' is 'rational animal'." Finally, we turn back to the individual human being, point to him and say: "How does he manage to fit in that logical classification which we have defined as 'rational animal'?" If we are sharp, we will say: "Because that man is a body which is animal, that man fits into the logical classification called 'animal'. Because that man is also a soul which is rational, that man fits into the logical *sub*-classification called 'rational'. That is how it comes about that, by

118

means of his body *and* his soul, that man is of the species called 'rational' of the genus called 'animal'."

Now then, point to an individual human being and ask: "How does he manage to fit in that logical classification which we have named 'substance'?" Aristotle will reply: "I define the logical classification 'substance' as 'that which is neither present in nor predicable of another'. Now then, that man there is matter and form. His matter is his substratum, and I define matter and substratum as 'that which is neither present in nor predicable of another'. Therefore, since this man is, in part, matter; and since matter is 'that which is neither present in nor predicable of another'—this man is, by means of his substratum called 'matter', included in the logical classification which we have termed 'substance'."

Point again to the same individual, and say to Aristotle: "How is he included in the classification called 'concrete individual substance'?"

I say that, in this case, Aristotle would reply: "You are speaking in a strained fashion. To speak of a classification is to speak of a universal—a characteristic common to a great number of individuals. The phrase, 'concrete individual substance', however, clearly indicates an individual rather than a universal. Nevertheless, if we must speak in your strained fashion, let us say this: I define 'concrete individual substance' as 'that which is composed of matter and form'. Since this man is matter and form, he is, by means of his matter and form taken together, included in the classification 'concrete individual substance'."

Such, then, is the difference between *defining things* and *defining definitions*. At all times, my dissertation was an examination into *the definition of definitions*, which is to say I was always concerned—*not* with real, individual things such as this man or this dead body—but, rather, with various descriptions of various general classifications.

Therefore, I considered the general classification called "substance", and asked how Aristotle defines this *logical substance*. I showed that he defines it as:

"THAT WHICH IS NEITHER PRESENT IN NOR PREDICABLE OF ANOTHER".

I then showed that Aristotle used the same phraseology to define the general classifications "substratum" and "matter". I showed further that, when speaking of the classifications called "substratum" and "matter", Aristotle also uses the phrases: (1) "potential substance", and (2) "that of which everything else is predicated, while it is not itself predicable of anything."

I then called for us to examine into "substance as substrata". In other words, having first shown that Aristotle used the same phrase (*i.e.:* "that which is neither present in nor predicable of another") in reference to both the general classification "substance" (*i.e.:* "*ousia*") and the general classification "substratum" (*i.e.:* "*hupostasis*"), I was now urging us to question to what extent that familiar phrase *should* apply to "substratum".

I went on to agree with Aristotle that *both* general classifications *should* each be described as "that of which all else is predicable while it is itself not predicable of anything else". I also agreed with him that the general classification "*ousia*" should *always* be *further* described as "that which is neither present in nor predicable of another". I disagreed with him, however, and said that, unlike the general classification "*ousia*", the general classification "*hupostasis*" *should not always* be *further* described as "that which is neither present in nor predicable of another". I suggested, instead, that it

120

was possible and better *sometimes* to describe it as "that which *is present in* another but *not predicable of* that other".

I then went on to show that, once we so *re*-describe the logical classification *"hupostasis"*, it is then possible to establish a general classification called "that of which all else is predicable, while it is itself not predicable of anything else", and to sub-divide it into these three general *sub*-classifications:

(1) **ACTUALLY** substantial substrata (*i.e.:* that which *actually is* neither present in nor predicable of another),

(2) **POTENTIALLY** substantial substrata (*i.e.:* that which *can become* that which is neither present in nor predicable of another),

(3) **INSUBSTANTIAL** substrata (*i.e.:* that which is always present in but never predicable of substantial substrata).

Next, I sought to show that, once we do that, we can then attach to the logical classification "matter" descriptions which are exceedingly different from those used by Aristotle to describe that logical classification. For him, the logical classification "matter" is described as "pure potency". For me, it is described as *neither* actuality *nor* potency; it is: "a line of force generated when an indivisible, non-spatial unit of actually substantial substrata acts in the presence of a drag inducing differential between itself and the indivisible, non-spatial, potentially substantial substratum to which it is hypostatically united". For Aristotle, the logical classification called "matter" has no application to anything in the real world, because, in Aristotle's conception of the real world, every real, individual thing is matter plus form. Thus, for Aristotle, the logical classification "matter" is *nothing but* a *logical* entity to be found only in the mind of man. For me, though—because *not all* substrata are "that which is neither present in nor predicable of another"—there is such a thing in the real world as a line of force which is not joined to a form, and, consequently, for me, the logical classification "matter" does have an application in the real world. For Aristotle, the logical classification "matter" applies to nothing which can be brought forward to account for anything that requires explanation in measurable terms. For me, though, it *does indeed* apply to such. Finally, for Aristotle, our sense images (*i.e.:* the shaped colors we see, the sounds we hear, the odors we smell, and so forth) are sensible forms stripped of all matter. For me, though, our sense images are both matter and form; consequently, when I see the color red I am face to face with what matter is being *in itself* in response to a form, which is to say I immediately intuit the internal characteristics which a definite number of lines of force are currently having in response to the particular form which is currently influencing them.

Regardless of appearances, none of the above statements is *primarily* an attempt to explain in what sense real things are substance, substratum, or matter. *Most of the time*, when I used the terms, "substance", "substratum", and "matter", I was *not* referring to *real* things. I was referring, instead, to *two expressions* taken directly from Aristotle, and the main thrust of all my efforts and words was to explain the two radically different ways in which those *two expressions* can be related to one another.

Those two expressions, of course, are:

(1)   that of which everything else is predicable, while it is itself not predicable of anything else,

AND

(2)   that which is neither present in nor predicable of another.

Having established the content of the two expressions, I showed that either one could equate the two (as Aristotle does) or that one could turn one into a genus and the other into a species of that genus (as Haas does), and I showed exactly how to do that.

After that *primary* goal was established, then—*and only then*—did I turn to real things in order to achieve a *secondary* goal. My secondary goal was to contrast the two radically different interpretations of reality which follow upon the two radically different ways of relating those two all-important expressions. Nevertheless, nowhere did I ever deny either:

(1)   that, for Aristotle, substance means far more than either substratum or homogeneous substratum,

(2)   that, for Aristotle, every primary individual substance is matter *plus form*,

(3)   that, for Aristotle, the matter of a real material body is "informed by some pair of properties".

SUBSTRATA

It is interesting to note again what your referee said in the quote which I used at the opening of section "C" of this letter. He said that much of the apparent confusion in Aristotle's treatment of *"ousia"* stems from the fact that ". . . in some contexts he is talking about definitions . . . whereas elsewhere he is concentrating on reality . . ." Even after mentioning the confusion that can follow from any reading of Aristotle *if one fails to distinguish between discussions of definitions and discussions of reality*, he then proceeds to make that very same mistake in his reading of my work. As a result, though I *repeatedly* said that, for Aristotle, each individual substance is matter plus form, he *still* hesitated not to assert that I was trying to say that every individual substance is, for Aristotle, nothing more than "homogeneous substratum". How strange it is!! If he had truly been mindful of the distinction between definitions and real things—a distinction to which he himself alluded—he could never have accused me of saying what he did accuse me of saying.

If my article has proven to be confusing to others, the reason for it is quite simple: Having read and understood Aristotle's method of speaking, I copied it together with its (as modern philosophers would now call it) confusing admixture of logic and ontology. I make no apology for what I did. I am a man with too many responsibilities—to the family of which I am a member and the community in which I live—to go searching hidden corners. I, therefore, copied Aristotle—one of the three greatest philosophers in the history of mankind and one whose works are both in the private book-collections of every thinking human being and on the more prominent bookshelves of every library in the world. My only "sin" is that I had not the time to search for the little known commentaries of unknown professors whose works can be found no place but buried in the lowest basements of no more than a dozen or so libraries.

(I)
CLOSING PLEA FOR FORGIVENESS:

Sir, if I have seemed a bit angry and insulting in this letter, I beg your kind indulgence. Twenty-two years of unremitting rejection and gross misinterpretation has left this forty-seven year old bachelor even more grouchy than old bachelors are expected to be. I am aware, of course, that my pains—however great they may be, either in fact or in fancy—do not *require* vehemence. It is always possible and better to quell wrath and to preserve courtesy. For that reason, if I *have* in fact been discourteous either to you or to your referee, then I offer you and him my profoundest apologies and humbly beg your forgiveness.

In closing, I again thank you for your unusually gracious gift and confess myself your eternal servant in return.

Very Truly Yours:

EDWARD N. HAAS

Before quitting the subject of freedom of opinion, it is fit to take some notice of those who say that the free expression of all opinions should be permitted, on condition that the manner be temperate, and do not pass the bounds of fair discussion. Much might be said on the impossibility of fixing where these supposed bounds are to be placed; for if the test be offence to those whose opinions are attacked, I think experience testifies that this offence is given whenever the attack is telling and powerful, and that every opponent who pushes them hard, and whom they find it difficult to answer, appears to them, if he shows any strong feeling on the subject, an intemperate opponent.

——**JOHN STUART MILL:** *Of Individuality* Chap. 2

℘art 2:

LETTER TO FR. CLARKE

Sunday June 12, 1983

Fr. Norris Clarke, S. J. - Editor,
"International Philosophical Quarterly,
Fordham University,
Bronx, New York. 10458.

Dear Father Clarke:

(A)
ACCUSED AGAIN OF SAYING THAT, FOR ARISTOTLE, SUBSTANCE IS NO MORE THAN SUBSTRATUM:

In your letter of rejection, dated June 7, 1983, you wrote as follows:

> My apologies for the delay in giving you an answer on your article on Aristotle and substance. . . . we do not find it fair to the full thought of Aristotle. . . . You do not seem to take into account the analogous

127

use of the term substratum, which can now refer to substance, not [sic! Did you mean "now"?] to what lies under change, and in substantial change this cannot be substance, evidently, since that changes, hence must be matter. Matter as substratum for subst. change is *part* of the whole substance, but not the whole.

Copies of my article, **Aristotle's Fundamental Fallacy**, went to every English-speaking philosophical publication on the North American continent and to the philosophy departments of over 30 universities in North America and Europe. In addition to your own, I received only three other commentaries in reply—one of which was no more than a small paragraph.

Judging by the above words taken from your letter, you have joined the other three in unanimously accusing me of having written an article on Aristotle's definition of *substance* (*i.e.:* "*ousia*" in the Greek). Like the others, you, too, seem to assert that, in my so-called "article on Aristotle and substance", I was trying to maintain that, for Aristotle, substance means nothing more than matter. I must say I am astonished to the utmost and utterly unable to understand how all of you could have made such a grossly tortured misinterpretation of my words.

The opening line of my dissertation was:

At issue here is one's basic ideas regarding substance as substrata. (pg. 241)

On page 248 of my paper, I wrote:

As "concrete thing", "substance" includes matter, form, and the composite of the two, and each of those three is also "substance".

On page 249 of my article, I wrote:

Though the following two quotes may not answer all our questions about what "substance" means to Aristotle, they at least tell us that matter is *potentially* substance and substratum without *actually* being such, and that explains how the matter *is* the substance in one sense while it *is not* the substance in another sense.

Finally, on page 250, I wrote:

Since, for the purposes of this dissertation, substance as substratum is our only concern, let us now ignore all the other meanings Aristotle attaches to "substance", and let us concern ourselves with what he says about substrata.

Am I perhaps so totally insane, that I cannot *truly* see what words I *actually* put down on the sheet of paper in front of me? Surely that must be the case. After all, to this altogether insane madman, it seems that I said—*in no uncertain terms whatsoever!*—that, for Aristotle, "substance" means far, far more than merely "substratum" or "matter"—that it means matter *and* form, and that, in meaning matter, it means only the *potential* part of the whole. To this hopelessly deranged lunatic, it seems that I then said quite plainly that we would ignore the multitude of *other* meanings attached by Aristotle to the word "substance", and would concentrate, instead, on only *one, highly restricted* meaning of "substance". In other words, I said we would consider the meaning of "substance" (*i.e.: "ousia"* in the Greek), **"ONLY IN SO FAR AS IT COINCIDES WITH THE MEANING OF 'SUBSTRATUM'"** (*i.e.: "hupostasis"* in the Greek).

To you, though—and to a great many others, apparently—such statements meant nothing save that I was trying to say "substance" (*i.e.: "ousia"*) means *nothing but* "substratum" (*i.e.: "hupostasis"*). It is as if a man were to say to a group of "scholars": "Now that we've shown that 'animal' can mean such things as lions and tigers as well as human beings, let us ignore all those other meanings and consider the term 'animal' only in so far as it means human beings;" and yet, *every* hearer were to leap to his feet and to shout indignantly: "How dare you say all animals are human beings? Are you so ignorant, you've never read any of the hundreds of scholarly works proving that lions, tigers, elephants, and bears too are animals?"

It absolutely *must* be that I am hopelessly psychotic and that, when I attempt to read the words I've written on a piece of paper, instead of seeing the words that are actually there, I see only words that are nowhere but in my own crazed mind. For, how else can I explain these misinterpretations so gross and so manifest as to defy both the capacity of the human *mind* to *believe* them and the capacity of the human *tongue* to *describe* them?

Over and over again I have asked myself if perhaps, in saying "substance as substratum", I had used some entirely new, wholly foreign phrase which no philosopher had ever encountered before—a phrase so strange to every eye and ear that none could possibly be expected to grasp the meaning of it. Is that somehow true? And yet, the phrase is found in Aristotle almost word for word. Thus, he writes:

> Since the substance which exists as underlying and as matter is generally recognized, and this is that which exists potentially, it remains for us to say what is the substance, in the sense of *actuality*, of sensible things.
> ——*Metaphysics*; Book VIII; Chap. 2; 1042b: 9. **Great Books**; Vol. 8; pg. 566 lower right.

Is it alright for Aristotle to speak of "substance as underlying and as matter" but wrong for me to speak of "substance as substratum"? Do you somehow understand *Aristotle* when *he* uses the phrase, but fail to understand *me* when *I* use it? If that is true, then it suggests to me a double standard hiding an ulterior motive, and I cannot help wondering, in that case, what I said that you people should find it so necessary to stop at nothing in your attempts to twist my words into meaningless gibberish.

On page 250 of my paper, I wrote:

> Very well, what *does* Aristotle say about substrata? I say that nowhere does Aristotle give us a thorough analysis of the idea of substrata. He says virtually nothing about it.

You and I both know that Aristotle says a great deal about the various meanings of "substance" ("*ousia*", if you wish). He devotes many words to "substance as actuality and form". But, if you know how much Aristotle says about the many meanings of "substance" and hear me say we are going to talk about something of which Aristotle says *virtually nothing*, how could you possibly understand me to be saying we would be talking about all the various meanings of "substance" and *"ousia"* in Aristotle?

I thought I made it exceedingly plain—both *from the beginning* of my article and *at many points in the body* of my article—that my article was about Aristotle and *substratum* (*i.e.:* "*hupostasis*")—a concept of which he says virtually nothing, because, in his estimation, that concept had already been sufficiently explored. Since my article was about Aristotle and *substratum* (and *not* "Aristotle and *substance*", as *you* said), then, in describing the main topic of my article, perhaps I should never have used the *phrase* "substance *as substratum*"; perhaps I should have always used only the *word* "substratum". In my estimation, though, the word "substance" ("*ousia*", if you wish) was a very desirable substitute for the excessively lengthy phrase: "that which is neither predicable of another, nor incapable of existence apart from another" (In a *slightly shorter* form, the phrase is: "that which is neither present in, nor predicable of, another".). Since my article was an examination into how *Aristotle did* apply that phrase to the concept of "substrata" ("*hupostasis*", if you wish) and how *we ought* to apply it, it seemed to me perfectly legitimate and logical to say my article was about "*substance* as substratum"; consequently, it seemed to me every reader—with even so little as the *slightest* background in philosophy—would easily follow my meaning.

Apparently, it was not as easy to follow me as I thought it would be. Still, to accuse me of saying that, for Aristotle, "substance" *never* means anything *other than* "substratum" is a new low in ludicrousness, since I *specifically* said we were concerned with "substance" (**i.e.:** "that which is neither predicable of another nor incapable of existence apart from another") *only in so far as* it applies to substratum. [16]

(B)
REJECT THE LIGHT, ## IF IT QUOTES NOT YOUR FAVORITES:

You wrote:

There have been many careful studies of substance in Aristotle, trying to unravel

---

[16] **NOTE OF MARCH 29, 2002:** In case the reader does not follow the above, my point was this: When I say that I am speaking of the term "substance" in so far as it applies to the term "substrata", it is merely an abbreviated way of saying that I am speaking of the phrase "that which is neither present in nor predicable of another" (**i.e.:** the phrase "that which is neither predicable of another, nor incapable of existence apart from another") in so far as it applies to the term "substrata". The great question is then this: Is it in fact correct to say that the phrase "that which is neither present in nor predicable of another" applies to <u>all</u> **substrata**? Aristotle, I contend, answers affirmatively. Indeed, I contend that *every* philosopher prior to myself answers affirmatively, and, in doing so, they all joined with Aristotle in promoting what constitutes Aristotle's chief fallacy. Indeed, they all thus joined in promoting what may well have been—for both Philosophy and Science—the most disastrous intellectual blunder in the history of the human race.

the obscurity in his treatment, but you
make no use of them.

In the first place, dear Father, what has that got
to do with anything? Was my article not long enough,
that I should have padded it with more quotes from
more sources?

But, you also wrote:

We find some interesting points in your
article.

If that is true, then you must have found some
light in what I said. Do you, then, reject the light because it fails to quote from the sources you like to see
quoted? All light is from God, and he who rejects the
light from God for so flimsy a reason as a lack of quotes
from authorities you like to see quoted—such a man, I
say, will answer to God for a grievous crime against the
light.

In the second place, Father, when you talk of
"careful studies of substance", what do you mean? In
what sense are they studies of substance? As Aristotle
says:

It follows, then, that 'substance'
has two senses, (A) the ultimate substratum, which is no longer predicated of
anything else, and (B) that which, being a
'this', is also separable, and of this nature
is the shape or form of each thing.
——*Metaphysics*; Book V; Chap. 8;
1017b: 24. **Great Books**; Vol. 8; Pg. 538
top right.

How many "careful studies" have you seen concerned with "substance" purely in the sense of "ultimate substratum"? If you know of any, I wish you would tell me where to find them. I have read St. Thomas, Maritain, Gilson, Anable, Bittle, Donceel, Dougherty, Dulles, Glenn, Joyce, Mc Cormick, Mercier, Phillips, Renard, Rickaby, Sertillanges, Van Steenberghen, Descartes, Spinoza, Locke, Berkeley, Hume, Bacon, Pascal, Kant, and Hegel among others, and I have yet to see a single, solitary analysis which, while concentrating upon the meaning of (as the Scholastics called it) "primary matter", discusses what I discussed.

I have a copy of a short treatise by Fr. Joseph Owens, C. Ss. R., entitled **Matter And Predication In Aristotle**, and it contains some good points, such as that, eventhough Aristotle removed all determinations and, hence, all intelligibility, from the "basic matter" of things (as Fr. Owens calls it), yet still he regarded the basic matter as, just in itself, a subject for predication.[17] Still, nowhere does even he discuss, as I did, the relationship, in Aristotle, between the two phrases:

(1)   that of which everything else is predicable, while it is not itself predicable of anything else,

AND

---

[17] **NOTE OF APRIL 22, 2002:** Fr. Owens' treatise is given as chapter 4 in **Aristotle, The Collected Papers Of Joseph Owens** as edited by John R. Catan and published by the State University of New York Press, Albany. Because that publisher's copyright notice is dated 1981, I assume the work was published that year. Turn to page 35 of that work, and read the first paragraph of the introduction.

(2)   that which is neither present in nor predicable of another.

In short, Father, I am by no means a man highly unfamiliar with what has been written, and I am familiar with nothing truly pertinent to what I wrote. If you knew of something pertinent, you could at least have had the common decency to give me the names of such works and could have told me where I might obtain copies of them.

(C)
TWO PHRASES OVER & OVER AGAIN:

Mine was not an (as you said) "article on Aristotle and substance". What, then, was it? It was an analysis of the applicability of the two phrases given above.

First of all, it was an analysis of the way in which Aristotle relates those two phrases, and what we saw is this: For Aristotle, *all* that is describable as:

(1)   that of which everything else is predicable, while it is not itself predicable of anything else,

is also describable as:

(2)   that which is *neither* present in *nor* predicable of another.

I showed that Aristotle applies both of the above phrases to both "substance" (*i.e.: "ousia"* in the Greek)

136

and "substratum" (*i.e.:* "*hupostasis*" in the Greek). For Aristotle, then, **NOTHING DESCRIBABLE AS**:

(1)   that of which everything else is predica-ble, while it is not itself predicable of any-thing else,

IS ALSO DESCRIBABLE AS:

(2)   that which *is present in* another but *not predicable of* that other.

But, one tires of repeating those two elaborate phrases over and over again. Therefore, for economy's sake, I substituted the word "substratum" for the phrase, "that of which everything else is predicable, while it is itself not predicable of anything else", and I substituted the word "substantial" (not the word "sub-stance") for the phrase "that which is *neither* present in *nor* predicable of another.

With such economizing definitions in hand, one is then able to say that, for Aristotle, *all* substrata are *substantial* substrata. But, to say that is merely to say— *in a short form!*—that, for Aristotle, *all* that is describ-able as:

(1)   that of which everything else is predica-ble, while it is not itself predicable of any-thing else,

is also describable as:

(2)   that which is neither present in nor predi-cable of another.

137

SUBSTRATA

To make that clarification was my first goal. Having attained that first goal, my second goal was to show that there is another way of relating those two phrases to one another. I suggested it is possible to say: *Not all* substrata are *substantial* substrata. That, of course, is but the *economical* way of saying: *Not all* that which is describable as:

(1)   that of which everything else is predicable, while it is not itself predicable of anything else,

is also describable as:

(2)   that which is *neither* present in *nor* predicable of another.

If *not all* substrata are *substantial* substrata, it follows that *some* substrata are _in_substantial substrata. That, of course, is but the *economical* way of saying: *Some* of that which is describable as:

(1)   that of which everything else is predicable, while it is not itself predicable of anything else,

is also describable as:

(2)   that which *is present in* substantial substrata *but not predicable of* substantial substrata.

In saying that, I departed dramatically from Aristotle's way of thinking and established a master classification to which I gave the name "substrata", and, in that master classification, I included *all* that which is describable as: "that of which everything else is predicable, while it is not itself predicable of anything else". That *master* classification includes these two *sub*-classifications:

(1)   whatever is *neither* present in *nor* predicable of another (*i.e.: substantial* substrata),

AND

(2)   whatever *is present in* substantial substrata *but not predicable of* substantial substrata (*i.e.: in*substantial substrata).

That *second* sub-classification admits of no further sub-divisions. That *first* sub-classification admits of these two further sub-divisions:

(1)   whatever *actually is* neither present in nor predicable of another (*i.e.: actually* substantial substrata),

AND

(2)   whatever *can be* neither present in nor predicable of another (*i.e.: potentially* substantial substrata).

I then went on to give you a technical explanation of how it was possible to have a kind of substratum which *is present in* substantial substrata *without* being *predicable of* substantial substrata. I introduced the concept of a line of force definable as: a line of tension formed when actually substantial substrata act in the presence of a drag inducing differential between themselves and potentially substantial substrata.

As I pointed out on page 280 of my paper, that definition of "lines of force" is *my* contribution to philosophy. It is a concept which neither Aristotle nor any other philosopher, save myself, ever grasped or even considered a possibility. Of this my claim to a unique, new idea, neither you nor anyone else ever said anything whatsoever. I wonder why.

(D)
WHY DISTINGUISH BETWEEN SUBSTANTIAL & INSUBSTANTIAL SUBSTRATA:

But, of what advantage is it to distinguish between *substantial* and *in*substantial substrata? In section "F" of my dissertation, I briefly alluded to the fact that, in making all substrata substantial substrata, Aristotle's system is wholly irreconcilable with quantum physics, since it does not allow one to think in the terms of either relative time or discrete quantities of matter and space. On the other hand, the idea that matter is *in*substantial substrata lays down the philosophical foundations for quantum physics.

Had I felt my paper could have afforded to be even more voluminous than it was, I would have pointed out something else, namely this: There are two radically different ways to conceive of the relationship between:

(1)   the world of what we see, hear, taste, touch, and smell (**i.e.:** the *actually* sensible world),

140

AND

(2)   the microscopic world of molecules, atoms, and sub-atomic particles (*i.e.:* the *potentially* sensible world).

He who says all substrata are *substantial* will conceive of that relationship one way, while he who says some substrata are *in*substantial will conceive of that relationship in a radically different way. The former will conceive of a relationship which puts an impenetrable wall between the *actually* sensible world and the *potentially* sensible world and, thereby, precipitates universal skepticism. The latter will conceive of a relationship which destroys that wall and which restores man's capacity to be certain. Let me explain why I say that.

What does it mean to say that all substrata are *substantial* substrata? Well, suppose Aristotle were alive today and were confronted with the demonstrable fact that this piece of paper can be broken down into a definite number of molecules each of which can be broken down into a definite number of atoms each of which can be broken down into a definite number of leptons and quarks each of which can be broken down into . . . . . .? We know not what. Were that the case, he, like virtually everyone else, would say: "It cannot be that this piece of paper is present in its molecules; it must be that its molecules are present in this piece of paper as potential parts of the whole. It cannot be that the molecules are present in their atoms; it must be that their atoms are present in the molecules as potential parts of the whole. It cannot be that the atoms are present in their leptons and quarks; it must be that the leptons and

141

quarks are present in their atoms as potential parts of the whole. It cannot be that the leptons and quarks are present in whatever comes next; it must be that whatever comes next is present in the leptons and quarks as potential parts of the whole."

Unfortunately, modern physics has irrefutably shown us two things: (1) It has shown us that we know nothing yet of the internal characteristics of any potential part of which we can *definitely* say that it is not merely another swarm of potential parts whirling around each other at unbelievable velocities in mind boggling patterns. (2) It has shown us that, when we examine the more basic potential parts of any object, those potential parts are—both individually and collectively—more and more radically different in themselves than is the whole in which they are present as potential parts. In short, it is exceedingly evident that—**_if_** it is true that the *actually* sensible world is *composed* of the *potentially* sensible world—then we *ultimately* know virtually nothing about the *fundamental* potential parts of anything, except that, whatever their internal characteristics might be, they are virtually nothing like what is given to our senses. **_But_**, if we know virtually nothing about the fundamental potential parts of this whole (except that they are radically different from what we see, hear, taste, touch, and smell), then how can we be sure that what we know about the whole is anything other than absolute nonsense? We cannot, and, for all we know, we *may well be* nothing more than dupes more totally deceived than we can even begin to imagine.

Therefore, if we say that all substrata are substantial, we deny that the matter of this piece of paper is a kind of substratum which is present in, but not predicable of, its molecules, and that means we are left with saying that it is, rather, the matter of the molecules which is being a potential part of the matter of this piece of paper. The minute that happens, we force ourselves into an intellectually suicidal regression—a regression attempting to explain the internal characteristics of the matter of this piece of paper in the terms of the internal characteristics of the matter of its molecules, which is, in turn, explained in the terms of the internal characteristics of the matter of its atoms, which is, in turn, explained in the terms of the internal characteristics of its leptons and quarks which, in turn, cannot be explained because we cannot yet be even reasonably certain that the leptons and quarks cannot be further reduced and, if reducible, cannot now determine to any extent whatsoever what the internal characteristics of those even smaller and more basic particles might be.

On the other hand, if matter is *in*substantial, then the molecules of this piece of paper are by no means parts of this piece of paper. Instead, the matter of this piece of paper is something more properly describable as "suspended between" the molecules, just as the matter of each molecule is something "suspended between" its atoms, just as the matter of each atom is something "suspended between" its leptons and quarks, just as the matter of each lepton and quark is something "suspended between" some more basic sub-atomic, material particle, until, at approximately 2^{-128} times the diameter of the hydrogen atom, there is a material object whose matter is (to use common sense parlance) "suspended between" the two "sides" of a unit of substantial substrata which cannot be physically split and which is without extension in space.

If the matter of this piece of paper is something "suspended between" its molecules, then the matter of this piece of paper is something which cannot have being if the molecules move too far apart. For that reason, the matter of this piece of paper is present in its molecules in the sense that it cannot have being apart from those molecules; and yet, the matter of this piece of paper is wholly distinct from the matter of its molecules, since the matter of each molecule is "suspended between" the matter of that molecule's atoms. That's why we say the matter of this piece of paper is *present in but not predicable of* its molecules.

To summarize, then, if we say matter is substantial, we are forced to look for the matter of this piece of paper *in* the very *internality* of its atoms, which means we are forced to look for the matter of this piece of paper *in* the very *internality* of its leptons and quarks, which means we are forced to look for the matter of this piece of paper *in* the very *internality* of . . . **WHAT**???!!! And thus is born an impossible regression which can only end in universal skepticism.

But, if we say matter is *in*substantial, we are forced to look for the matter of this piece of paper *in between* the individual molecules in this piece of paper, and that means our search, if properly conducted, will end right there, and what we will see is this: What is *in between* the molecules is, when taken as a whole, exactly like what we see, hear, taste, touch, and smell.

(E)
ACCUSED OF INFINITE REGRESSION:

You wrote:

Also if anything exists, substance must exist. For if everything is in something else, we have an infinite regress; somewhere there must be something that exists in itself and not as part of another, or nothing could ever exist at all.

Good grief, Father! How much *more* pathetic can this affair become? Are you *seriously* trying to accuse me of having said that *everything* is **_in_** something else? I spent twenty pages **FIRST**: distinguishing between: (#1) "that which is neither present in nor predicable of another", (#2) "that which is present in, but not predicable of, another"; and **SECOND** in showing how #2 is present in, and generated by, #1. After all of that, do you now propose to accuse me of asserting that there is no #1???!!! Surely you jest!

To be sure, I said *all matter* is that kind of substratum which is *in something else*; nevertheless, rather than saying that *everything* is *in something else*, I even went so far as to give you a mathematical description of the point at which we pass **FROM**:

(#1) that which *is present in* something else,

TO:

(#2) that which *is not present in* something else.

Thus, on pages 269 & 270 of my article, I wrote:

. . . substrata can be divided only to approximately 2^{-256} times the diameter of the hydrogen atom[18] before one reaches

---

[18] **NOTE OF APRIL 8, 2002:** The above ought to read: "substrata can be divided only to approximately 2^{-128} times the diameter of the hydrogen atom". Why, then, did I use the figure 2^{-256} instead of 2^{-128}? Twenty years ago, I was still thinking in the terms of what I then called the "diminished alphatopon".

an indivisible unit of substrata. . . every ultimate, indivisible unit of substrata is, as *substantial* substratum, intrinsically di-une, having its potency as a form of negative reality only logically separable from the positive reality of its actuality.

Since I had defined *substantial* substrata as "that which *is not present in* something else, the above quote *necessarily* meant this: If, at this point in time and space, you try to find any *material* object smaller than approximately .1 trigintillionths of a centimeter (*i.e.:* 1/10 000 000 000 000 000 000 000 000 000 000 000 000 000 000 000 000 000 000 000 000 000 000 000 000 000 000 000 000 000 cm.) in diameter, you will pass *from* the three-dimensional world of "that which *is present in* another" (*i.e.:* the world of matter) and *into* the non-spatial world of "that which *is not present in* something else". No doubt, though, you will still persist in accusing me of saying that *everything* is *in something else*. May God have pity on us all!

(F)
"IN ANOTHER" DOESN'T ALWAYS

As I *then* saw things, it was only in *most* cases true that the smallest unit of matter in the universe was 2^{-128} times the diameter of the hydrogen atom. Still, in *some* (if not *many*) cases, that smallest unit could be reduced to the square of that figure, which is to say 2^{-256} times the diameter of the hydrogen atom. Where that is true, then, it is necessarily true that, if we are looking for the diameter below which one can *never* find divisible substrata (*i.e.:* matter), then that dia. is 2^{-256} times the diameter of the hydrogen atom.

146

MEAN "PART OF ANOTHER", UNLESS YOU'RE AN ARISTOTELIAN:

Consider again that last set of words which I took from your letter. To repeat, you said:

> Also if anything exists, substance must exist. For if everything is in something else, we have an infinite regress; somewhere there must be something that exists in itself and not as part of another, or nothing could ever exist at all."

Why are you so concerned, dear Father, with (to use your words) "something that exists in itself and not as a part of another"? You seem to be implying that whatever does *not* exist *in itself* is necessarily *part of another*. What makes you think there cannot be something which, while it is <u>in</u> another, *is* <u>not</u> either a *part* of that other or something *predicable* of that other? I cannot help suspecting, Father, that you too are guilty of the same fundamental fallacy which held Aristotle in its grip. As a result, you too, like Aristotle, fail to understand the full meaning of the words: "substance", "substratum", "being", "subject", "thing", and "independence".

All of you Aristotelians seem to have a *compulsive* need to describe both "substance" and "substratum" as **BOTH**:

(1)   that of which all else is predicable, while it is not itself predicable of anything else,

AND

147

(2)   that which is neither present in nor predicable of another.

The result invariably is that you irrevocably
blind yourselves to two-thirds of the meaning of each
one of the words: "substratum", "being", "subject",
"thing", and "independence". Why do I say that?

Before proceeding, for the sake of economy, let
us say this: The phrase, "that of which everything else is
predicated, while it is itself not predicable of anything
else", shall be taken to be equivalent to the term "subject" (We have long ago said it is equivalent to "substratum".). This is appropriate because that which is a subject is the subject of a predicate. It is that of which the
predicate is predicated and, consequently, is not itself
the predicate which is predicated. The term "subject"
shall also be taken to be equivalent to the term "non-
predicable". Every subject is a non-predicable, and
every non-predicable is a subject. The reason for that
should be self-evident.

Following Aristotle, the phrase "present in another" means the same as the phrase "incapable of existence apart from that other". Again for the sake of
economy, let us say this: The phrase "inhering in another" shall be equivalent to the phrase "present in another in the sense of 'incapable of existence apart from
that other'". The term "inhering thing" shall mean: "that
which _is_ so present in another as to be incapable of existence apart from that other". The term "_non_-inhering
thing" shall mean: "that which is _not_ so present in another as to be incapable of existence apart from that
other".[19]

---

[19] **NOTE OF APRIL 8, 2002:** In other words, "inhere in another" means to be so dependent upon *cohering* to another
that annihilation is unavoidable without that cohering to an-

Here, then, is **_the_** great question which precipitates all the confusion: *Is every subject a non-inhering subject?* To put it another way: Is *every* non-predicable a *non-inhering non-predicable?* Aristotle's answer has always been _yes_. For him, and for all other Aristotelians, every subject is a non-inhering subject, which is to say every subject is a substantial subject, and "subject" and "substantial subject" are equivalent terms. Unfortunately, the minute you Aristotelians say that, you destroy philosophy.

Aristotle apparently loves phrases like this one: "in the strictest, most proper, and unqualified sense". I like such phrases too, so let's use it.

We can now say this: Because you Aristotelians say all non-predicables are *non-inhering* non-predicables, you *necessarily* conclude that such terms as "substratum", "thing", "being", "subject", and "independence", each have—*when taken in the strictest, most proper, and unqualified sense*—only *one* meaning, which is to say each signifies what is "*sui generis*", which is to say each indicates a genus which admits of only one species. Needless to say, that one species is: "that which is neither present in nor predicable of another". In short, all such terms—*when taken in the strictest, most proper, and unqualified sense*—apply only to substance. At least, so it is in the Aristotelian system.

Some will perhaps challenge that and argue that, for Aristotle, "substance", "substratum", "being", and "thing" have *two* meanings. After all, does not Aristotle speak of both *potential* and *actual* substance, *potential* and *actual* being, and so forth?

---

other. "Inhering thing" thus means "that which so coheres to another as to be incapable of avoiding annihilation unless it continues to cohere to (*i.e.:* continues to be intimately bonded to) that other".

SUBSTRATA

That is true. For Aristotle, though, potential substance, and potential being are merely logical entities. They are in no sense real things, which is to say they are, for Aristotle, never a subject capable of halting the gaze of the intellect. They are, rather, devoid of all determinations and all direct intelligibility.

To explain what that means, contrast what "potential being" means to Aristotle and what it means to me. Asked what it means to him, there is nothing to which Aristotle can point and say: "That's what I mean by 'potential being'". Ask me, however, what "potential being" means, and I will point to empty space. Empty space is not, as some would have it, a sheer vacuity of all forms of being (*i.e.:* being in a strict, most proper, and unqualified sense). It is, instead, something having being in a strict, proper, and unqualified sense of the word "being". As such, it is a subject and a kind of substratum.

So much so is that true, that, as a substratum, it has internal characteristics, and its most important internal characteristic is its diameter. The diameter of space is so real a thing that it can be varied at will by the observer. That is to say that the diameter of space is relative to the quantity of actuality being actualized by the observer. What does that mean? Since it is an exceedingly technical issue, I will not get into it in this letter.

To say "empty space" is a *real* thing called "potential being" means this: When you or I take cognizance of the empty space around us, each of us is being immediately aware of his own undeveloped "pure potency" presenting itself to our gaze exactly as it is in itself as a substratum. That pure potency, however, is *not* a *part* of you. It is another subject and another substratum which is inextricably *bonded to* your substratum

150

but still standing *outside of* your substratum. Nevertheless, you *can*, so to speak, "drink it up" and *convert* it into a part of you. It is, so to speak, *your* auxiliary fuel tank bolted to you—to you the rocket-ship to eternity—and it is waiting for you to command it to pour its fuel into your rocket engines. That's why it stands immediately adjacent to you and in a profoundly intimate union with you. It is, so to speak, ever pressing against you and begging to be taken in and converted into a part of you. In death, you *will* answer its plea, pull it in totally, and become as vast in yourself as the universe now appears to be to your eyes. (One might say that the pre-eminent distinction between man and the other animals is that man can be aware of empty space as a subject while the other animals can not.)

To the unthinking eye, it seems that we are "in space". In fact, though, it is space which is "in" each of us, if by "in us" you mean, not as parts are in a whole, but "incapable of having being apart from us". Destroy all the objects "in" space or convert all those objects into "pure actuality", and there will be no space, because the *potentially* substantial substrata which were being space will either have joined their substantial hosts in oblivion, or they will have been converted into pure actuality.

Such, then, is what it means *to me* to say that "potential substance", "potential being", and "matter" have more than one meaning *even when taken in the strictest, most proper, and unqualified sense*. Certain it is that, for Aristotle, those terms could *never* have *real* meaning in the same sense that they have *real* meaning for me. As many a writer has pointed out, of course, that has always been *the* great problem with Aristotelianism: Except when it talks about "concrete substance", it never gets around to indicating *any* direct object of experience. It's all words indicating other words which

151

indicate other words *"ad infinitum"*. You can never pin-point the meaning of anything in Aristotle, because it's all abstractions and concepts floating around in an utterly mental world which never touches base with the real world except in the fleeting moment when it points to a man and a horse and says: "That is a concrete substance composed of matter and form." Even then, though, neither Aristotle nor anyone else ever knew then, or knows now, what either matter or form *really is*. All he can say is that it's something *like* this and something *like* that, but it's not *really* this or that. It's all analogy, analogy, analogy, with not a single, solitary *real* idea *anywhere* in the thousands of pages of words, words, words.

Having said all of that, though, I must, nevertheless admit with admiration that, although he worked with nothing but analogies and words, Aristotle still managed to lay down—*for the most part*—the logical boundaries within which the real thing would have to fall when it was finally found. The only place he went wrong was in thinking that *whatever* is describable as:

(1)　that of which everything else is predicable, while it is not itself predicable of anything else,

is also describable as:

(2)　that which is neither present in nor predicable of another.

To get back on the track, I repeat: For you Aristotelians, "substratum", "being", "thing", "subject", and "independence"—*when taken in the strictest, most proper, and unqualified sense*—have only *one* meaning in the *real* world. For me, though, those words—*even when taken in the strictest, most proper, and unqualified sense*—have *three* meanings *even* in the *real* world. Thus, just as "animal"—*even when taken in the strictest, most proper, and unqualified sense*—can mean either we rational animals or the irrational animals around us, so also does "substratum"—*even when taken in the strictest, most proper, and unqualified sense*—signify either *actually* substantial substratum (*i.e.:* that which cannot be sensed), *potentially* substantial substratum (*i.e.:* space), or *in*substantial substratum (*i.e.:* the lines of force which we experience as colors, shapes, odors, flavors, sounds, etc.).[20] By the same token, "subject"—*even when taken in the strictest, most proper, and unqualified sense*—signifies either inhering subjects, actually non-inhering subjects, or potentially non-inhering subjects.

(G)
INDEPENDENCE VS.
NOT STUCK TOGETHER:

In all probability, some Aristotelians will object: "An inhering subject is not an independent subject. But, a subject cannot be a subject unless it is independent; therefore, there can be no such thing as an independent, inhering subject. By its very nature, it is a self-contradiction and, hence, an impossibility."

---

[20] We are not aware of space by means of our sense images. That is to say we do not *sense* or *feel* space. We are aware of space by means of our souls, which is to say we are aware of space by means of a form of experience which is exceedingly different from our bodies' sense perceptions. Contrary to what many wish to believe, we *do indeed* have more than *one* way to experience the world around us.

I reply that "independence" and "physical separation" are not equivalent terms. Independence does not *necessarily* mean the ability to stand apart from another. *Perfect* independence no doubt implies such, but *not all* independence is *perfect* independence. The "*sine qua non*" of independence is "non-predicability", and whatever has "non-predicability" *truly* has independence *in some strict, proper, and unqualified sense of the word "independence"*.

Consider, for example, the case of Siamese twins so extensively joined that they cannot be separated without killing one or both of them. They most certainly are not as independent of one another as you and I are independent of one another; nevertheless, each is an individual person with an independent will, personality, and soul.

Come to think of it, just how independent *are* you and I from one another? Without the earth beneath our feet, the atmosphere around us, and other people to grow our food, weave our clothing, and produce the building materials for our houses, how long do you suppose we would continue to live? Independence, therefore, admits of degrees. But, as long as there is "non-predicability" there *truly* is independence *in a strict, proper, and unqualified sense of that word*. In short, this is just another example of you Aristotelians trying to force *one* meaning upon a word which necessarily has *three* meanings in the real world.

And what about "being"? For you Aristotelians, all "being" is "existence". Is that the way it *must* be? I say it is not.

As you undoubtedly know, Father, the word "existence" comes from the Latin word "*existo*", which means: "I stand outside of." Existence, therefore, is that kind of being characteristic of that kind of subject which stands outside of all other subjects. It is the being of "that which is neither present in nor predicable of another". It is the being of "non-inhering non-predicables", which is to say it is the being of "non-inhering subjects".

What about "inhering subjects"? Do they "exist"? Do they "stand outside of" all other subjects? By the very meaning of their names, they do not; therefore, their being is not existence.

Therefore, whereas, for you Aristotelians, all *real* being (*i.e.:* being in the strictest, most proper, and unqualified sense of the word "being") is existence (because, for you, *potential* existence is but an abstraction), for me, *real* being is either *actual* existence, *potential* existence, or *in*existence. By "Inexistence", I mean the *real* being of that which can never *exist* but which must always be so present in another as to be incapable of having being apart from that other.

Naturally, "inexistence" can be translated as "non-existence" and "non-being". Contrary to what men tend to think, "non-existence" and "non-being" do not indicate a sheer vacuity of all *real* forms of being. They indicate, rather, the presence of an unusual but nevertheless *real* form of being.

For me, "actual existence" is equivalent to each of these two terms: "actually existential being" and "actual existentiality"; and "potential existence" is equivalent to each of these two terms: "potentially existential being" and "potential existentiality"; and "inexistence" is equivalent to each of these two terms: "non-existential being" and "non-existentiality".

If some *real* being (*i.e.:* being in the strictest etc.) is *non*-existential being; and if, in seeing this piece of paper, I see such *real* being—then why should I worry about regressing to existence at all? I have already attained to *real* being and true knowledge of all that this

155

piece of paper is *in itself*. Therefore, certitude about the veracity of my knowledge of the *real* being of horses and the internal characteristics of their matter is by no means what requires me to regress in search of existence and substantial substrata. What requires me to worry about existence and substantial substrata is my need to gain so perfect a control over all horses, that I can then guarantee myself that no horse will ever kick my brains out. In other words, exhaustive knowledge of *existence* gives exhaustive knowledge of *causality*, and exhaustive knowledge of causality gives exhaustive power over the elements of my environment, and exhaustive power over the elements of my environment gives exhaustive power to live a long, healthy, happy life. In short, if some substrata are *in*substantial and some being is *non*-existential, then it is *self-preservation* and *not* universal skepticism which requires me to find existentiality and that ever elusive kind of substratum which is "neither present in nor predicable of another".

(H)
TIME FOR INSULTS:

It is my unique contribution to philosophy that I am the first human being in history to work out in exact detail the relationship between substantial and insubstantial substrata and between existential and non-existential being. Indeed, I have even gone so far as to work out—to a somewhat large extent—the mathematical laws (based on an intrinsically triune infinity) governing the point in time and space at which insubstantial substrata finally reduce no further and leave us at long last in the bosom of substantial substrata. Concerning this claim of mine, however, none of you says anything whatsoever. Why is that? I suggest, dear priest, that the answer is this: You are face to face with ideas too advanced either for you or for any but the smallest few of your fellow university type philosophers.

The vast majority of you university type Aristotelians are dead men entombed forever in a dead past. That's why, every time the modern world commences to speculate about the internal characteristics of *substrata*, something compels *you* to seal your ears and to run away screaming: "They're trying to equate substance with substratum and to say every substance is nothing more than its sub-atomic particles."

When will you understand, O blind guides of the blind? The vast majority of modern thinkers have never sought to make any such equation, precisely because the meaning of "substance" is both *meaningless* to them and *of no interest* to them. It is considered a waste of time by them, because they have already *fully comprehended* what you people have been saying over and over again, namely: that such terms have, for philosophy, no *real* meaning; they are only analogies having nothing more than an *analogous* meaning and nowhere pointing to anything observable. But, unlike you university type philosophers, most men's minds cannot be content with similitude and pure abstractions with no concrete counterpart in the real world. After all, we *live* in the *real* world and *not* in a *dream* world of amorphous ideas which never do anything but approximate what we see, hear, taste, touch, and smell. You cannot feed your children with something that is "like" food but isn't *really* food; you can't clothe your children with something that is "like" clothing but isn't *really* clothing; and you can't shelter your children with something that is "like" wood but isn't really wood. That's why the vast majority of modern thought has merely shifted the focal point of its attention from deductions about the meaning of "substance", "matter", and "form" to an inductive analysis of the internal characteristics of whatever can in any way be measured either directly by the senses themselves or indirectly by means of instruments which can be read by the senses. More importantly, though, it has done such with virtually no concern whatsoever for the abstract questions concerning either the meaning of substance as a "category" or the meaning of substance as "the substance of a thing". Inquiry into the characteristics of substrata is precisely that—an inquiry into the characteristics of *substrata* abstracted and separated from all other questions.

You the dead philosophers are the *only* ones making all the loud fuss about equating substratum with substance, because it is *you* who cannot tell the difference between a discussion about substance and a discussion about substratum. That is why, though I said quite plainly that mine was a discussion *deliberately ignoring* the more expansive meanings of "substance" and *deliberately limiting* itself to possible interpretations of "substratum", virtually every one of you more than forty university types jumped in to call my article an article on substance—an article, all of you insisted, attempting to analyze the *full* meaning of "substance" in Aristotle and, thereby to prove that, for Aristotle, "substance" meant no more than "homogeneous substratum". Consequently, your main concern was with utterly irrelevant, pedantic, and sophomoric questions about why I didn't sprinkle my text with the original Greek terms and quotes from "the commentators"—as if a work on what *Aristotle* said should contain mainly the words of someone *other* than Aristotle.

When that many people respond with the same ridiculously blind, and pathetically gross misinterpretation of what another man has said, it clearly bespeaks, in the hearers, a neurotic preoccupation with, and readiness to overact to, certain key phrases which—because of the neurotic reaction which they automatically provoke in the ears of the neurotic hearers—make it impossible for those hearers to take note of anything else which might have been said.

Whence comes this obviously demented reaction? I will tell you where it comes from, Father. It comes from an overly sensitive ego which has lost the capacity to defend itself in a constructive manner. What makes me say that? I will explain.

For over two thousand years, a veritable ocean of philosophers labored at the art of deductive reasoning in an attempt to bring forth the light of truth to the human race. Though that great a number labored at the task for that many years, they failed even so much as to suspect what the scientists, laboring at the art of _in_ductive reasoning, discovered "overnight" (comparatively speaking) in the infancy of science and the inductive method.[21]

---

[21] **NOTE OF APRIL 8, 2002:** For some, "induction" _primarily_ means drawing into the field of attention (whether that of your own self or that of others or both) some number of _observed_ particulars as grounds for a general assertion. For others, it means any reasoning in which the conclusion does not _necessarily_ follow from the premises no matter how well they support the conclusion. For those in the _former_ category, "induction" conjures up images of the _scientist_ laboriously applying all kinds of instruments to the observation and measurement of nature's phenomena, while the _philosopher_ is forever seated in his armchair laboriously studying definitions. For those in the _latter_ category, "induction" conjures up images of the _critically_ thinking _philosopher_ demanding _absolute certitude_, while the _un_critically thinking _scientist_ is content with what's merely _well evidenced_. In the above, as I spoke of "inductive reasoning" and the "inductive method", I was speaking as one of those in the _former_ category.

SUBSTRATA

You university type philosophers have become accustomed to saying that the philosophers of old made no attempt to conclude what could or could not be found in the microscopic world within the objects which our senses grasp. In my opinion, you are exceedingly wrong, but it is unimportant. Regardless of whether or not you are correct, the undeniable fact is that the vast majority of their educated and thinking contemporaries believed that, in their philosophers, they had found ideal guides to the exact nature of what we sense. Certain it is, too, that they understood those philosophers to be telling them that the world hidden from view could not possibly be as the scientists suddenly proved it to be by throwing it up to view.[22]

---

[22] **NOTE OF APRIL 8, 2002:** What a shock it was to the Aristotelians when Galileo's telescope revealed that the surface of the moon is not perfectly smooth when brought into close view! Many of them were so affected, they refused to look, lest they see for themselves how wrong they had been. How totally ignorant of history one would have to be to imagine philosophy sought not to predict the properties of things not readily observable! Sadly, to preserve their egos, the *pseudo*-philosophers—rather than admit philosophy tried it and fell on its face miserably—do imagine it and daily make even greater fools of themselves with a flood of pretentious dissertations in which they seriously try to deny philosophy's colossal flop. As for myself, I have taken a different tack: I admit the flop but, clinging to hope in philosophy's abilities, I shall make a new attempt to uncover even the most fundamental laws of the universe and to do it by merely sitting in my armchair and laboriously toying with my own mental activities and the definitions which my own and the mental activities of others have devised. Time will tell whether I have merely repeated the flop or exonerated philosophy so totally as to leave all the world awestruck forever.

Suddenly, the vast majority of those with any significant degree of either education or thinking ability found themselves convinced that—in achieving as much as it did as swiftly as it did—science had irrefutably proven that philosophy in general, and metaphysics in particular, are, at best, a fool's errand—at worst, a vicious attempt to perpetuate human ignorance in order that the ruling class might hold their fellow men in bondage forever. Predictably, in the eyes of the vast majority of capable men, to be a devotee of metaphysics was suddenly to be the consummate "social dropout" wasting one's time and talents in one of the most disgraceful ways possible to human beings. In short, even today, in the eyes of virtually every intelligent and supposedly intelligent man, to be a metaphysician is to be no less an object of contempt than a professional thief is.[23]

All we human beings have egos, and it is not easy for any ego to cope with that much contempt from the world around it. Every attempt to deal with it will almost certainly result in an extremely sensitive ego. The constructive man, eventhough he may have an extremely sensitive ego, manages to overcome it all by truly finding the ability to convince himself that he *will* one day win the approval which he deserves. Unfortunately, far too many of those few who go on pursuing metaphysics have taken a different tack. In response to that great a quantity of contempt and ridicule, they have found that the best response is an equal amount of

---

[23] **NOTE OF APRIL 22, 2002:** If memory serves me correctly, I once read somewhere that there is no Nobel Prize for philosophy, and that's because Alfred Nobel declined to set one up—declined specifically because he felt no one should be awarded anything for engaging in a pursuit as totally worthless as philosophy is.

contempt and ridicule. I, for one, can truthfully say that, in my experience, the most bitterly contemptuous people I have ever met have always been those who fancy themselves "serious philosophers".

If I go to a doctor and ask him to heal my sore throat, he prides himself on being able to help me without my having to learn medicine. If I go to an attorney and ask him to defend me against an accuser, he prides himself on being able to help me without my having to study law. If I go to an accountant and ask him to help me with my tax forms, he prides himself on being able to help me without my having to learn bookkeeping. But, if I go to a philosopher and ask him to heal my universal skepticism, he merely smiles condescendingly when he hears my thoughts, tells me the problem is merely that I don't know what I'm talking about, and then tells me to go study philosophy.

For every specialist but the "serious philosopher", the sign of success is his ability to convey the fruits of his specialty to others without those others having to learn his specialty. For the "serious philosopher", though, what makes him so special is the fact that those outside of his specialty cannot enjoy the fruits of his specialty. The "serious philosophers", therefore, ever dwell in a vacuum where they speak only to fellow "serious philosophers". How they recognize each other is quite simple: They know what Greek terms to drop and quote liberally from the latest favorites. And, if perhaps, once in a while, they should wonder how it could come to pass that the world could admire the scientists much more than it admires them, they quickly remind themselves that it's because the world is too lazy and too ignorant to study and to master those angelic, analogous definitions which only the great masters of words know.

It is no wonder for me, that the mass of humanity has rejected philosophy. For, none but a fool would give any attention to a profession which prides itself on its lack of usefulness to everyone outside of that profession.

Lest you misunderstand me, Father, let me hasten to point out the difference between *true* contemptuousness and only *apparent* contemptuousness. Some of us will sometimes screw up our faces in anger and give the other guy a piece of our minds, because we are convinced there's a strong possibility that a hefty, verbal kick in his pants will help him to see straighter. Still, we have no delusions about ourselves as we do so. We know that it is highly unlikely that we are any better than he is. We know that we, too, could probably use a hefty tongue-lashing. If it seems to us we see straighter than he does for the moment on some particular issue, we know only too well how quickly that situation may reverse itself.

There are others, though, whose chief objective is to let the other guy know that they are superior to him and his contemptible mannerisms. Such men generally never show the slightest bit of anger. After all, they are above such "animalistic" tendencies so common to the "immature" individuals around them. Poor blind fools! They imagine that because they have castrated themselves they have become angels, when, in reality, all they've done is to turn themselves into freaks.

In the past three hundred years, the truly contemptuous "serious philosophers" have developed a very neat little "ego defense mechanism" which allows them to maintain their sense of disdainful superiority over the scientists in a very easy way. They have become experts in proving that the modern world has completely failed to understand the meaning of "substance" and "matter" because it has never grasped the esoteric art of "meaning by analogy". As a result, there has now come about a kind of ritualistic routine which every "serious philosopher" immediately tosses off the minute he's confronted with anyone who mentions the key word "substratum". After all, that's the word which conjures up those terrifying memories of what science did to philosophy with the microscope. Anyhow, the minute he hears that key word (*i.e.:* buzz word, if you prefer), the "serious philosopher" *must* abruptly terminate the conversation. He *must* then smile condescendingly, pat the poor, misguided ignoramus on the head as he reminds him of how little he understands about the *true* meaning of substance, and then tell him to learn Greek and read for another twenty years. It would perhaps be more infuriating were it not so pathetic.

You are no doubt, Father, saying to yourself: "What has brought on this outburst?" Perhaps you have forgotten what you wrote:

> Also the tone of discussion is a bit irreverent or flippant at times, not quite the scholarly dispassionateness one would look for in this erudite area.

You may rest assured, my dear Father Clarke, that I am most certainly indeed a flippant, irreverent, and passionate man. I am a forty-seven year old man with nothing more than a high school diploma. For me, philosophy is but the hobby of an uneducated amateur. I am also, however, the lord and master of assets in seven figures, and I did not get where I am by being either a fool or a pusillanimous wimp. I am well known and hated by many of the local politicians and lawyers, and our scrapes have been on the front pages of the local newspapers on several occasions. In short, Father, whatever I have achieved in life—*in whatever field of endeavor*—I have achieved only because I am a born fighter and, unlike you, God has loved and co-operated with my pugnacity.

"God loves and gives grace to the pugnacious?!" you will *probably* gasp in objection. Yes, I know how it *tends* to be with people in the universities: They think only gentle little lambs can be saved or serve as great instruments of God's light. For them, all the lions are worthless and damned, because *anything* which roars—*for whatever reason it roars*—is intrinsically evil (or "sick" as they would prefer to put it). But, let me tell you something, dear Father. I have read Aristotle several times, and I dare say you are well aware of how flippant and irreverent Aristotle was. I have read the New Testament from cover to cover at least fifty times, and I can assure you that even our Lord Jesus Christ was exceedingly flippant, irreverent, and passionate, and it was difficult for St. Paul to open his mouth without provoking a riot. I have read most of the Fathers and Doctors of the Church (in the original Latin wherever available), and I can assure you that many of them were quite flippant, irreverent, and passionate. Louis Pasteur was so flippant, irreverent, and passionate, that he had many a shouting match on the floor of the French Academy of Medicine, and, on at least one occasion, engaged the eighty-year-old surgeon general of France in a fist fight. You will perhaps pardon me, therefore, if I make bold to say I like the company I keep and am not the least intimidated by your contempt for passionate words.

Indeed, O priest of God, I will wax even more flippant, irreverent, and passionate, and I will tell you this: I have mixed with, and been considered one of, some of the most successful men in my part of the world, and it has been my experience that truly great, learned, and successful men couldn't care less whether or not you are flippant, irreverent, or passionate. Do you know why they don't care? It's because everyone of them knows that he, too, is the same way and that he likes being that way. It is only children, *truly* contemptuous adults, unduly sensitive egomaniacs, and the mentally ill who take issue with you the minute your tone of vice veers one iota from that of a "Liberace", because they are silly enough to believe that they themselves are above such behavior.

I am not disturbed, O priest of God, by your charge of insolence. What *does* pique me, however, is this: that you *initiated* the charge. When another *initiates* the charge of flippancy, irreverence, and passionateness, he is not merely charging you with flippancy, etc.; he is also saying: "The reason I can make such a charge, of course, is that I am not myself guilty of such sick behavior. I am above these personality defects which so obviously plague you." In short, O priest of God, you have not so much accused me of being flippant and irreverent as you have accused me of being inferior to you in virtue, perfection of character and merit before God and man. I suggest to you, O priest of God, that that is *real* flippancy and irreverence and that, in initiating such an attack, you have proven yourself to be not quite as untainted as you think you are.

I admit again that I am indeed flippant, irreverent and passionate. But, such *human* emotions are by no means *intrinsically* evil. As long as they are directed only at individuals who deserve a good thrashing, they are naturally good, and the man who engages in such activity is "guilty" of nothing more than being less than angelic or divine. On the other hand, if a flippant, irreverent, and passionate man goes about *persistently* trying to convince himself and others that he is *above* flippancy, irreverence and passionateness, such a man is guilty of being a liar, a zealous hater of truth, a devil, and a man destined for eternal damnation.

Because I am a Roman Catholic and you a priest, Father, I am extremely reluctant to rail at you this way. But, as St. Paul did not hesitate to call St. Peter a hypocrite (**Galatians** 2: 11) when the circumstances made it fair play to do so, so also will I not hesitate to tell you to your face that you have injured me gravely without just cause. You have thrown into my teeth grievous insults which are totally out of proportion. Had I, in my article, unjustly accused anyone of "sick behavior", you would have had the right to denounce me, and I would have no reason now to complain. *But*, there was nothing whatsoever in my article to merit that you should treat me in so contemptuous a manner. Were you somehow offended by my assertion that Aristotle's notions regarding substance and substrata are "nothing but totally gratuitous assumptions"? In that case, I give you the words of St. Paul:

> To the pure, all things are pure, but to the corrupt and unbelieving, nothing is pure. Their very minds and consciences are corrupted.
> ——*Titus*; 1: 15

How is it flippant or irreverent or passionate to tell a man he has *some* totally gratuitous assumptions? I for one am not so stupid as even to begin to assert that I am free of all gratuitous assumptions. And how much more stupid would I have to be to take offense at the one who brings those gratuitous assumptions to my attention???!!!

As you can see by reading this letter, if I *intend* to be flippant, irreverent or passionate, I leave no doubt about it in *anyone's* mind. I, you see, do nothing half-heartedly; I do everything *full throttle*. When I wrote **Aristotle's Fundamental Fallacy**, I took exceedingly great pains to make sure that there was nothing acrimonious in that work. But, just as you twisted from my words a meaning which is diametrically opposed to what I actually said, so also did you pervert the spirit behind that work into the very opposite of what *it* was. Neither the meaning nor the motive which you read into my work was in the mind of the writer; it was nowhere but in the mind of the reader.

Still, suppose I *had* been flippant, irreverent, and passionate—what of it?! The important point is: Was I flippant to one who had done nothing to deserve the degree of flippancy I directed at him? Was I irreverent to one who had done nothing to deserve the degree of irreverence I directed at him? That issue, you did not even raise. You did not question whether or not my flippancy, irreverence, and passionateness were *unfair* retribution. The flippancy, irreverence, and passionateness which you read into my work, you rejected out of hand without any concern for whether or not it was fair play.

What it all boils down to, Father, is this: In the *first* place, the light which God sent you through me, you rejected on the grounds that it dared to come to you clothed in the skin of a lion—a creature whose ways are, in your eyes, intrinsically unacceptable to your world. In the *second* place, the lion whose ways you contemned was *you—not me*. *Therefore*, not only did you abuse me with a charge which I had done *nothing* to merit, *but*, rather, you dared also to denounce in me a characteristic which is by no means intrinsically evil and which is manifestly as much your characteristic as it is mine.

Perhaps it seems to you, Father, that I am now being overly abusive of you. If I am, then let a disinterested party or parties judge the case. I will, if God will but co-operate with my efforts, always submit to the judgment of Rome, because it is the judgment of the universal church that such is the way all genuine Catholics should act. Still, remember this: You, Father, started this business of insulting a man for daring to be a human being rather than an angel, and I am sick and tired of you so-called "dispassionate scholars" cursing my human nature. When I get finished with you, you will never again dismiss the writings of anyone on the grounds that they are irreverent, flippant or "not quite the scholarly dispassionateness one would look for in this erudite area".

Let me assure you, my dear Father Clarke, that my work *will* be published, because I am wealthy enough to see to it. It is merely a question of finding the time to attend to it. Pray, Father, that my thoughts are merely nonsense which will go unnoticed by the world, because, if they prove to be otherwise, I assure you that you will not like it. I am saving these insulting rejection letters from you and your fellow *non*-irreverent, *non*-flippant, *dis*passionate scholars, and, if and when the world comes to my door, I will make you and your fellow "dispassionate scholars" eat those letters word by word, and I guarantee you I will make you the laughing stock of this planet.

(I)
AN OLIVE BRANCH TO CLOSE:

Now that my vehement side has emptied its wrath, dear Father, let another side of me have its say.

SUBSTRATA

It is part and parcel of being a passionate man to feel real pain whenever one hears that another human being—no matter how remotely related to one's self—has experienced pain. For that reason, I am truly pained to hear that you have just come out of major surgery. May God grant that, in hearing of my pain over your pain, you will experience at least a small diminution in your own suffering.

Let us pray to God for one another and thank Him most heartily for both the pleasure and the displeasure of this chance meeting in the midst of our brief sojourn upon the shores of this land of death.

Sincerely,

EDWARD N. HAAS

Part 3:

LETTER TO MR. BIGGER

Tuesday September 20, 1983

Mr. Charles P. Bigger
Dept. Of Philosophy
Louisiana State University
Baton Rouge, La. 70803

Dear Mr. Bigger:

So great is the quantity of time which has lapsed since you wrote to me, it is quite probable that you no longer remember having done so. In case such has in fact occurred, I have taken the precaution of including a copy of your letter to me.

My principal purpose in writing to you is to extend to you the courtesy of thanking you for your generosity. Of the thirty universities to which I sent copies of my paper, ***Aristotle's Fundamental Fallacy***, not a one bothered to send me a single, solitary word, except for Louisiana State University, and Oxford University in England. As brief as your letter is, it consequently stands out as a remarkably prominent gesture of common decency and concern for one's fellow man. Seeing

171

how unique your act has now manifested itself to be, I am so deeply touched, that I find myself wondering if there are words which can adequately express my gratitude.

On the *outside* chance that you *might* be interested in the reaction of Oxford University to my essay, I am, under separate cover, sending you a copy of a letter from a Mr. Theodore Scaltsas, B.A., M.A., D.Phil. Together with it, I am including a copy of my reply to his letter. I am sending these items under separate cover and forewarning you of it so that, if, as is quite probable, you do not wish to be bothered, you can, without even opening it, recognize a nuisance you wished to avoid and easily toss it in the trashcan.

Incidentally, I did in fact also send copies of my paper to every English-speaking, Philosophy journal on the North American continent. Only one stated it was worth publication, though it could not use it. Most sent only "canned" rejection slips. Three sent extremely derogatory and insulting replies.

Gratefully yours:

EDWARD N. HAAS

P.S.: Mr. Scaltsas is apparently the only one who realized that my paper was an analysis of Aristotle's interpretation of the concept of *substratum* and had virtually nothing to do with either *substance* or *being* or *"ousia"* as developed by Aristotle.

$\mathfrak{P}$art 4:

LETTER TO MR. SCALTSAS

Mr. Theodore Scaltsas
Department Of Philosophy
New College
Oxford, England
OX1 3BN

Dear Mr. Scaltsas:

To begin with, please accept my heartfelt thanks for your letter. Of the thirty universities to which I sent copies of my paper, *Aristotle's Fundamental Fallacy*, only one other university, besides your own, bothered to reply, and the other response contained no criticism whatsoever.

Having said that, I will now proceed to impose upon your time even further with a detailed reply to your few, very valuable remarks.

(1)

SUBSTRATA

You are quite correct. I ought not to say "sub-strata is". Why I used the singular verb with the plural noun, I do not know. I can only say that I goofed, and I am certainly very grateful that you brought this rather silly error to my attention. I ought to have caught it my-self. [24]

(2)

You wrote as follows:

> . . . you introduce too many terms . . .
> which you do not take the time to explain
> and elaborate at length. . . . it would help
> the reader very much if you spent more

---

[24] **NOTE OF MARCH 30, 2002:** Whether or not a noun is plural depends upon how you use it. For example, if I'm using the noun "animals" to refer to one or more of the many _extra_-mental creatures to which it is most often used to refer, then the noun is plural, and one ought to say "animals _are_", as in the statement "This zoo's animals are always asleep." Oppo-sitely, if I'm speaking of the noun itself, then it's singular, and one ought to say "animals _is_", as in the statement "The word 'animals' is used to refer to an immense range of targets in the world around us." Likewise, if I'm using the noun "ani-mals" to refer to the _intra_-mental concept we attach to each of those targets, then, again, the noun is singular and one ought to say "animals _is_" as in the statement: "'Animals' is a concept composed of several other concepts, including that of self-movement." My use of "substrata _is_" should have been a "dead give-away"—both to myself and particularly to men with university degrees—that I was mainly speaking of the word and the concept it signifies rather than of the many things in the extra-mental world which "substrata" is rightly used to signify. Unfortunately, like myself, the "alphabet soup" people were too dense to catch it at the moment.

space in giving examples of what you mean.

In response to that very informative piece of advice, I offer the following outline:

DEFINITIONS

I. ARISTOTELIAN FOUNDATIONS :

i. Now the substratum is that of which everything else is predicated, while it is itself not predicated of anything else.
——*Metaphysics*; Book VII; Chap. 3; 1028b: 35. **Great Books Of The Western World**; Encyclopedia Britannica Inc.; Chicago, 1952. Vol. 8: pg. 551 bottom left & top right.

ii. We have now outlined the nature of substance, showing that it is that which is not predicated of a stratum, but of which all else is predicated.
——*Metaphysics*; Book VII; Chap. 3; 1029a: 6. **Great Books**; Vol. 8: pg. 551 upper right.

iii. Substance, in the truest and primary and most definite sense of the word, is that which is neither predicable of a subject nor present in a subject; for instance, the individual man or horse.
——*Categories*; Chap. 5; 2a: 11. **Great Books**; Vol. 8: pg. 6 left just above the middle.

iv.   By being 'present in a subject' I do not mean present as parts are present in a whole, but being incapable of existence apart from the said subject.
——*Categories*; Chap. 2; 1a: 22-23. **Great Books**; Vol. 8: pg. 5 lower left.

v.   . . . in the truest and primary and most definite sense of the word . . .
——*Categories*; Chap. 5; 2a: 11. **Great Books**; Vol. 8: pg. 6 left just above the middle.

vi.   Further, primary substances are most properly so called, because they underlie and are the subjects of everything else.
——*Categories*; Chap. 5; 2b: 38. **Great Books**; Vol. 8: pg. 7 left just above the middle.

vii.   Everything except primary substances is either predicable of a primary substance or present in a primary substance.
——*Categories*; Chap. 5; 2a: 34. **Great Books**; Vol. 8: pg. 6 near top right.

viii.   Moreover, primary substances are most properly called substances in virtue of the fact that they are the entities which underlie everything else, and that everything else is either predicated of them or present in them.
——*Categories*; Chap. 5; 2b: 15. **Great Books**; Vol. 8: pg. 6 right just below the middle.

ix.   It follows, then, that 'substance' has two senses, (A) the ultimate substratum, which is no longer predicated of anything else, and (B) that which,

being a 'this', is also separable, and of this nature
is the shape or form of each thing.
——*Metaphysics*; Book V; Chap. 8; 1017b: 24.
Great Books; Vol. 8: pg. 538 top right.

x. Since the substance which exists as underlying
and as matter is generally recognized, and this is
that which exists potentially, it remains for us to
say what is the substance, in the sense of *actuality*, of sensible things.
——*Metaphysics*; Book VIII; Chap. 2; 1042b: 9.
Great Books; Vol. 8: pg. 566 lower right.

xi. By a 'primary' substance I mean one which does
not imply the presence of something in something else, i.e.: in something that underlies it.
——*Metaphysics*; Book VII; Chap. 11; 1037b: 2.
Great Books; Vol. 8: pg. 561 upper left.[25]

---

[25] The above quote succinctly states the principal difference between myself and Aristotle. For *me*, every three dimensional "primary substance" *does* imply "the presence of something in something else". "This man" or "this horse" implies (*in addition to a form*) a spatial substratum which *must* of *necessity* be "present in something else". Because matter is "that which is no longer predicated of anything else" (see quote ix above), matter is "the ultimate substratum" of "this man", "this horse", and any other individual, three dimensional thing. For all of that, contrary to the above assertion from Aristotle (one he repeats many times; see quote ii above), matter *must* be *present in* another and more independent kind of substratum. By "present in", I mean, like Aristotle (see quote iv above), that the ultimate substratum called "matter" is "being incapable of existence apart from the said subject". The "said subject", in which matter must be present, is, in my system, the more independent kind of ultimate substrata called "the generators of matter".

177

xii.   It is plain from what has been said that both the name and the definition of the predicate must be predicable of the subject. . . . . With regard, on the other hand, to those things which are present in a subject, it is generally the case that neither their name nor their definition is predicable of that in which they are present. Though, however, the definition is never predicable, there is nothing in certain cases to prevent the name being used. For instance, 'white' being present in a body is predicated of that in which it is present, for a body is called white: the definition, however, of the colour 'white' is never predicable of the body.
——*Categories*; Chap. 5; 2a: 19 & 26. **Great Books**; Vol. 8: pg. 6 left middle.

xiii. It is just that we should be grateful, not only to those with whose views we may agree, but also to those who have expressed more superficial views; for these also contributed something, by developing before us the powers of thought.
——*Metaphysics*; Book II; Chap. 1; 993b: 11. **Great Books**; Vol. 8: pgs. 511-512.[26]

---

[26] In listing the above quotes, I am not by any means attempting to assert that, for Aristotle, "substance" and "substratum" are equivalent terms. I'm only saying that, for Aristotle, as for virtually all thinking individuals, "substance" *sometimes* means the same as "substratum", and—*in so far as* they *do* mean the same—they mean "that which is neither predicable of, nor so present in, another as to be incapable of existing apart from that other".

II. PRELIMINARY CLARIFICATIONS :

A. "To *inhere in* a subject" = "to be so present in a subject as to be, at least here and now, incapable of existence apart from *intimate bonding to* that subject".

B. "*Substantial*" = "*non*-inhering"; whereas, "*in*substantial" = "*inhering*". In other words, in this system, "inhering" shall always mean "being present in a subject" and, like Aristotle, "by being 'present in a subject', I do not mean present as parts are present in a whole, but being incapable of existence apart from the said subject" (see quote iv above, and remember that, as I see it, in using "apart from", Aristotle, as with myself, meant "apart from *intimate bonding to*"). Naturally, "non-inhering" means <u>not</u> "being present in a subject". By "<u>not</u> being present in a subject", I mean <u>not</u> being incapable of existence apart from *intimate bonding* to a subject. Repeat that distinction between "substantial" and "insubstantial". Repeat it to yourself a thousand million times. It is <u>*the*</u> all important key to all that follows.[27]

---

[27] **NOTE OF MARCH 30, 2002:** The phrase "incapable of existence apart from another" is greatly subject to misunderstanding. When <u>*I*</u> speak of a first something as incapable of existence *apart* from a second something, I mean the first something cannot exist unless it is *intimately bonded* to the second something. To illustrate what I mean (and I contend Aristotle meant the same thing), imagine Siamese twins are so extensively united, it's medically impossible to separate them without killing them both. In such a circumstance, it

C. "To be *part* of a subject" = "to be so present in a subject as to be able, whether in the present or in the future, to exist apart from that subject" = "to be present in a subject without inhering in it".

D. "That which is inhering in a subject" = "*ens in altero*" (**CAUTION**: In this system, "accident" and "*ens in altero*" are not equivalent terms.).

E. "That which does not inhere in a subject" = "*ens in se*".

F. In this system, every *thing* is either: (1) God, (2) an angel, (3) a human soul, (4) a form, (5) a generator, (6) the undeveloped potency of a form or

---

would be quite correct to say that neither is capable of existence "apart from" the other. In this case, though, "apart from" means neither is capable of existence without being *intimately bonded* to the other. Oppositely, it can also and rightly be said that neither of them is capable of existence "apart from" a certain narrow range of temperatures, pressures, air content, food, and so forth. In that latter case, however, "apart from" doesn't even come remotely close to meaning what it means in the former case. On the face of it, there is simply no comparison between: (1) the way in which the twins are *intimately bonded* to one another, and (2) the way in which the twins are *accompanied* by the required environmental conditions. To put it yet another way, there is no comparison between the way my body's *shape* is connected to my body and the way my body is connected to this planet. A properly equipped rocket ship can break the connection between my body and Earth and do so without any harm to either. On the other hand, break the connection between my body and its current shape, and the shape, at least, will cease to exist.

generator, (7) a line of force generated by a generator (or group of lines of force generated by a
group of generators), or (8) a characteristic predicable of one or more of those six types of subjects.

G.   In this system, God, the angels, human souls,
forms, and generators are each non-spatial subjects, which is to say they are *not* three-dimensional, which is to say they do not occupy space.
God, the angels, human souls, and forms are also
"incorporeal". To say they are "incorporeal" is to
say they do not generate lines of force within
themselves. Generators, while being non-spatial,
are corporeal, which is to say they *do* generate
lines of force in themselves. Exactly how generators _do_—while God, the angels, the forms, and
human souls do _not_—generate lines of force is not
within the scope of this literary effort.

III. "SUBJECT" = (*i.e.:* is equivalent to each of
the following:)

i.   "that of which everything else is predicated,
while it is itself not predicated of anything else";
ii.   "a non-predicable";
iii.   "a substratum";
iv.   "a thing in the *primary* sense".

EXAMPLES: God; the angels; the human souls;
the forms; the generators; the potency of a generator or a
form; lines of force.

A. "SUBSTANTIAL SUBJECT" =

    i.   "a substantial non-predicable";

    ii.   "a non-predicable which does or can exist actually separated apart from all other substantial non-predicables";

    iii.   "that species of the genus 'non-predicables' which does not inhere in another non-predicable";

    iv.   "that kind of subject which is _not_ so present in another subject as to be incapable of existence apart from that subject";

    v.   "that which is neither inhering in, nor predicable of, anything else";

    vi.   "a non-parasitic non-predicable".

EXAMPLES: (1) **ACTUALLY** substantial subjects: God; individual angels; individual human souls; individual forms; and the individual non-spatial, indivisible generators which generate sensible matter; (2) **POTENTIALLY** substantial subjects: the undeveloped potency of a generator or of a form.

B. "INSUBSTANTIAL SUBJECT" =

    i.   "an insubstantial non-predicable";

    ii.   "a non-predicable which must always inhere in another non-predicable";

    iii.   "that species of the genus 'non-predicables' which can never exist separated apart from all other non-predicables";

    iv.   "that kind of subject which _is_ so present in another subject as to be incapable of existence apart from that subject";

v.   "that which *is never* predicable of anything else, but which *is always* inhering in another";

vi.   "a parasitic non-predicable".

EXAMPLES: matter; every spatially material object including we humans though only from the standpoint of each person's body.

IV. "THING" =

i.   "that which is";
ii.   "that which has being";
iii. *"quod est"*.

A. "SUBJECTIVE THING" =

i.   "that which is immediately present to the consciousness of an intelligent subject;"

ii.   "that which is either a sense image or an idea of which an intelligent subject is currently being aware;"

iii.   "that which has being either partly or entirely within a field of consciousness;"

iv.   "a thing of consciousness;"

v.   "a thing of awareness;"

vi.   "a mental thing;"

vii. "an encountered;"

viii. "that which is either a direct or a reflex encountered;"

ix.   "a thing only in a manner of speaking."

EXAMPLES: names; definitions; categories; classes; genera; species; differentiae; sights; sounds; flavors; odors; tactile feelings; concepts; ideas; acts of will; figments; illusions; etc.

1. "PARTLY SUBJECTIVE THING" =

i. "a partly mental thing;"
ii. "a predicable thing;"
iii. "a mental thing having a foundation in an extra-mental reality;"
iv. "a thing of consciousness legitimately predicable of an extra-mental thing;"
v. "an internal representation of an external reality;"
vi. "that which is either an idea or a sense image having a foundation in an extra-mental reality;"
vii. "that which is either an idea or a sense image legitimately predicable of an extra-mental thing;"
viii. "a non-imaginary encountered;"
ix. "a direct or reflex encountered which, it is legitimately proposed, correctly describes extra-mental things."

EXAMPLES: names, definitions, concepts, ideas, categories, classes, genera, species, differentiae, and universals which, it is legitimately proposed, correctly describe extra-mental things; sights, sounds, flavors, odors, and tactile sensations involuntarily induced by contact with extra-mental things, which is to say sense images which, it is legitimately proposed, correctly reproduce in the mind, material things which exist outside the mind.

a. "PREDICABLE IDEA" =

i. *"species expressa;"*
ii. "a predicable thing stripped of all sensation;"
iii. "a partly mental thing abstracted from all material content;"
iv. "an immaterial predicable thing;"
v. "an objective idea;"
vi. "an objective concept;"
vii. "a non-imaginary idea;"
viii. "a non-imaginary concept;"
ix. "a non-imaginary reflex encountered;"
x. "a reflex encountered predicable of an extra-mental thing."

EXAMPLES: names, definitions, categories, classes, genera, species, ideas, concepts, differentiae, universals, and all other kinds of abstract classifications which, it is legitimately proposed, correctly describe extra-mental things.

(1) "ACCIDENTAL IDEA" =

i. "an immaterial predicable thing concerning the quantity, quality, relation, place, time, position, state, actions, or affection of a subject;"
ii. "an objective idea legitimately predicable of the accidental characteristics of subjects;"

iii. "an objective concept whose name and definition are *not both* predicable of a subject;" (see quote xii from Aristotle)

iv. "a reflex encountered correctly describing the temporary characteristics of a subject."

EXAMPLES: names, definitions, categories, classes, genera, species, all other kinds of classifications, concepts, and ideas which, it is legitimately proposed, correctly describe what subjects are accidentally.

(2) "NON-ACCIDENTAL IDEA" =

i. "a mental thing legitimately predicable of a subject *as a subject;*"

ii. "an objective idea concerning the non-accidental characteristics of subjects;"

iii. "an objective concept concerning that which is legitimately predicable of, but not inhering in, a subject;"

iv. "an idea concerning the indispensable characteristics of subjects;"

v. "what subjects are per se to the viewer's mind;"

vi. "an idea whose name *and* definition *are both* predicable of a subject." (quote xii from Aristotle)

EXAMPLES: names, definitions, categories, classes, genera, species, all other kinds of classifications, concepts, and ideas which, it is legitimately proposed, correctly describe what makes subjects subjects.

b. "PREDICABLE PHANTASM" =

    i.   *"species impressa;"*
    ii.   "a sense image predicable of an extra-mental subject;"
    iii. "a sensible, mental thing which, it is legitimately proposed, reproduces in the mind the accidental characteristics of some particular, extra-mental subject;"
    iv. "a non-imaginary direct encountered."

EXAMPLES: any sight, sound, odor, flavor, or tactile sensation involuntarily induced by contact with an extra-mental subject, which is to say any sense image which, it is legitimately proposed, is the result of an extra-mental subject impressing upon the mind a reasonably true copy of one or more of that extra-mental subject's characteristics.

2. "PURELY MENTAL THING" =

    i.   "an entirely mental thing;"
    ii.   "a purely imaginary idea or sense image;"
    iii. "a purely imaginary reflex or direct encountered;"
    iv. "a mental thing which is not legitimately predicable of any extra-mental thing;"

187

  v. "a chimera;"
  vi. "a figment."

EXAMPLES: mermaids; satyrs; square circles; mercy killing.

B. "OBJECTIVE THING" =

  i. "thing absolutely;"
  ii. "thing strictly speaking;"
  iii. "thing;"
  iv. "that which is either a subject or a characteristic of a subject."

EXAMPLES: God; Angels; souls; forms; undeveloped potency of a generator or a form; the generators; lines of force; all material objects; the quantity, quality, relation, place, time, position, state, action, or affection of the preceding.

1. "THING IN THE SECONDARY SENSE" =

  i. "a predicable";
  ii. "an accident";
  iii. "that which is always both inhering in, and predicable of, a subject";
  iv. "a non-subject".

EXAMPLES: any color, shape, quality, quantity, relation, and so forth of a subject.

2. "THING IN THE PRIMARY SENSE" =

i.   "a non-predicable";
ii.   "a subject";
iii. "a stratum";
iv.   "that of which everything else is predicated, while it is itself not predicated of anything else".

EXAMPLES: matter; the generators of matter; the undeveloped potency of a generator or a form; forms; the human soul; the human body; any and all material objects; the angels; God.

a. **"INDEPENDENT THING IN THE PRIMARY SENSE"** =

i.   "a substantial thing in the primary sense";
ii.   "a substantial subject";
iii. "a subject which does not inhere in another;"
iv.   "that which is BOTH undivided in itself AND either actually or potentially divided apart from all other substantial subjects";
v.   "a substantial hypostasis";
vi.   "an uncompounded hypostasis";

EXAMPLES: (1) **ACTUALLY** independent things: each of the astronomical number of forms and generators of matter; individual souls; individual angels; God; (2) **POTENTIALLY** independent things: the undeveloped potency of a generator or of a form.

b. **"IMPERFECTLY INDEPENDENT THING IN THE PRIMARY SENSE"** =

i.   "an insubstantial thing in the primary sense";
ii. "an insubstantial subject";
iii. "that which is undivided in itself but can never be divided apart from *all* other subjects";
iv. "that which is undivided in itself but must always inhere in some substantial subject";
v.   "that which is undivided in itself *and also* either actually or potentially divided apart from all other subjects *governed by the same species of time-space form*"; (The reader is not expected to understand the meaning of that.)
vi. "an insubstantial hypostasis";
vii. "a compound hypostasis";

EXAMPLES: matter (*i.e.:* each of the lines of force generated by the generators of matter); each and every spatially material object including man, though *not* from the standpoint of his soul (Though it is abnormal for the human soul to exist apart from its body, it is, at any time, able to do so. It is, therefore, an independent thing in the primary sense of the word "thing".).

V. "ESSENCE" =

i.   "what a thing is";
ii. "that which in some sense individuates a thing";

iii.   *"quid quod est".*[28]

A. "ESSENCE IN THE SECONDARY SENSE" =

   i.   "that which individuates a thing in the secondary sense";

   ii.   "that which individuates a predicable";

   iii. "that which individuates an accident";

   iv. "that which individuates a non-separate thing as a non-separate thing".

EXAMPLES: essence of any color, shape, quality, quantity and so forth.[29]

B. "ESSENCE IN THE PRIMARY SENSE" =

   i.   "that which individuates a subject";

   ii.   "that which individuates a thing which is in some legitimate sense separate";

   iii. "that which individuates a non-predicable";

---

[28] *Exactly* what "essence" *ought* to mean, I cannot truly determine. It sometimes means the sum total of the characteristics (internal and otherwise) of a thing; other times means only the absolutely indispensable characteristics; and sometimes means the absolutely indispensable causes of a thing.

[29] When "essence" is used in this sense it is more _ab_used than *used*.

EXAMPLES: essence of God, an angel, a human soul, a form, a generator, the potency of a generator or a form, or lines of force.

1. "SUBSTANTIAL ESSENCE IN THE PRIMARY SENSE" =

i. "that which individuates a substantial subject";
ii. "that which individuates an independent thing in the primary sense";

EXAMPLES: (1) **ACTUALLY** substantial essence: essence of a generator's actuality; essence of a form's actuality; essence of a human soul's actuality; essence of an angel's actuality; essence of God; (2) **POTENTIALLY** substantial essence: essence of the undeveloped potency of either a generator or a form.

2. "INSUBSTANTIAL ESSENCE IN THE PRIMARY SENSE" =

i. "that which individuates an insubstantial subject";
ii. "that which individuates an imperfectly independent thing in the primary sense";

EXAMPLES: essence of matter (*i.e.:* essence of the lines of force generated by the generators); essence of a horse; essence of a human body; essence of any material object.[30]

---

[30] **NOTE OF APRIL 22, 2002:** These days, I would further distinguish between *logical* and *ontological* essence. "Ontological essence" refers to the sum total of the internal characteristics of some particular subject or its accidents. All the above are ontological essences. "Logical essence" refers to whatever serves as the crucial indicator by means of which an intelligent observer determines precisely how to fit—within his chosen system of classification—that which he wishes to classify. Ontological essences are *particular*; logical essences are not, because, by their very nature, they are designed to tell us in what way two or more ontological essences are similar. Logical essences are often founded on ontological essences, but not always. Sometimes, we classify readily observable ontological essences on the basis of the not so readily observable behavior of the molecules and/or atoms which underlie them. Since such behavior is difficult to observe, the crucial indicator becomes the effect of one readily observable ontological essence upon another such. For example, trying to determine whether some readily observable ontological essence should be classified as either an acid or a base, the crucial indicator becomes its effect on litmus paper. By the same token, it's easy enough to determine whether or not some life form is a biped: All you have to do is to look at its ontological essence and see how many legs it includes. But, to determine whether or not it is a *rational* animal, we cannot merely look to see if the ontological essence of its body is bonded to the ontological essence of an immortal soul. All we can do is to watch for a length of time and see if it sometimes acts on the world around it the way only rational life forms can. Only the observ*ed* determines what is *truly* its *ontological* essence; but, the *logical* essence is purely relative to the observ*er*'s chosen system of classification—a choice very much dependent upon the observer's objectives, education, and so forth.

VI. "BEING" includes:

A. "BEING IN THE SECONDARY SENSE", which pertains to:

    i.   all things in the secondary sense;
    ii.   all predicables;
    iii. all accidents;
    iv. all essences in the secondary sense;

EXAMPLES: being of any color, shape, quality, quantity, and so forth.

B. "BEING IN THE PRIMARY SENSE" = "BEING OF A SUBJECT" :

1. "SUBSTANTIAL BEING IN THE PRIMARY SENSE" pertains to:

    i.   all independent things in the primary sense;
    ii.   all substantial things in the primary sense;
    iii. all non-parasitic non-predicables;
    iv. all substantial hypostases;

EXAMPLES: Actually substantial being in the primary sense pertains to God, the angels, human souls, the forms, and the generators. Potentially substantial being pertains to the undeveloped potency of a generator or of a form.

2. **"INSUBSTANTIAL BEING IN THE PRIMARY SENSE" pertains to:**

    i.   all dependent things in the primary sense;

    ii.   all insubstantial things in the primary sense;

    iii. all parasitic non-predicables;

    iv. all insubstantial hypostases;

    v.   all inhering subjects.

EXAMPLES: Insubstantial being in the primary sense pertains to all lines of force and all material objects but only from the standpoint of their matter.

VII. "SUBSTRATUM" = "THAT OF WHICH EVERYTHING ELSE IS PREDICATED WHILE IT IS IT-SELF NOT PREDICATED OF ANYTHING ELSE". It includes:

A. **"SUBSTANTIAL SUBSTRATUM" = "ANY NON- PREDICABLE WHICH EITHER CAN BE IN THE FUTURE, OR CURRENTLY IS, NOT INHERING IN ANOTHER NON-PREDICABLE". It includes:**

1. **"ACTUALLY SUBSTANTIAL SUB-STRATUM"** = "that which *actually is* neither predicable of, nor inhering in, any other subject". It includes:

   i. God;
   ii. the angels;
   iii. the human souls;
   iv. the generators;
   v. the forms

2. **"POTENTIALLY SUBSTANTIAL SUB-STRATUM"** = "that which *can become* that which actually is neither predicable of, nor inhering in, any other subject". It includes:

   i. the potency of a generator or a form.

B. **"INSUBSTANTIAL SUBSTRATUM" = "THAT WHICH IS NEVER PREDICABLE OF, BUT MUST ALWAYS INHERE IN, A SUB-STANTIAL SUBSTRATUM".[31] It in-cludes:**

---

[31] Each insubstantial substratum is a line of tension formed between the actuality and the potency of some substantial sub-stratum as that generator acts in the presence of a drag in-ducing differential between its actuality and the inhering, un-developed potency.

i. all lines of force (*i.e.:* every piece of matter);
ii. every spatially material object (because the substratum of every three-dimensional object is a definite number of lines of force).

VIII. "ENTIA" =

i. "beings";
ii. "whatever has being in any sense";
iii. "anything which *is* in any sense";

A. "*ENS IN ALTERO*" =

i. "being in another";
ii. "thing in another";
iii. "that which must inhere in a subject";

EXAMPLES: lines of force; potency of a generator or of a form; colors; shapes; quantities; qualities; etc.

1. "THING EXISTENTIALLY PRESENT IN ANOTHER" =

i. "*ens in altero secundum quid*";
ii. "that which has *only* its *existence* in another";
iii. "subject inhering in another subject";
iv. "that which, while it cannot have being unless it inheres in a substantial subject, nevertheless has internal characteristics which are not part of the internal characteristics of the subject in which it inheres";

    v.   "that which inheres in another but which is not predicable of that other";

    v.   "that which has both being and essence in the primary sense, but whose being is insubstantial";

EXAMPLES: potency of a generator or of a form and lines of force.

    a.   **"THING ALWAYS EXISTENTIALLY PRESENT IN ANOTHER" = "lines of force".**

    b.   **"THING SOMETIMES EXISTENTIALLY PRESENT IN ANOTHER" = "the potency of a generator or form".**

2.   "THING ESSENTIALLY PRESENT IN ANOTHER" =

    i.   *"ens in altero simpliciter"*;
    ii.   "accident";
    iii.   "that which has both its being and its essence in another";
    iv.   "that which has both being and essence only in the secondary sense";
    v.   "that which has no internal characteristics save those which are part of the internal characteristics of the subject in which it inheres";
    vi.   "that which both inheres in a subject and is predicated of that subject.

EXAMPLES: colors; shapes; quantities; qualities; and so forth.

B. *"ENS IN SE"* =

i. "being in itself";
ii. "thing in itself";
iii. "that which does not inhere in another";
iv. "that which has its being in itself";
v. "that which stands alone";
vi. "that which can continue to exist even without being intimately bonded to another";
vii. "that which exists in the truest, primary, and most definite sense of the word 'exist'";
viii. "that which is involved in activity after the manner of an agent."

EXAMPLES: God; the angels; the human souls; the forms; the generators.

1. *"ENS ACTUALITER IN SE"* =

i. "an *actual* being in itself";
ii. "an *actual* thing in itself";
iii. "that which is already not inhering in another";
iv. "that which is already standing alone";
v. "that which *is* existing";
vi. "that which *is* involved in activity after the manner of an agent."

EXAMPLES: God; the angels; the human souls; the forms; the generators.

a. *"ENS ACTUALITER IN SE A SE"* =

i. "that which caused itself to be an actual thing in itself";
ii. "an uncreated, actually substantial subject";
iii. "God".

b. *"ENS ACTUALITER IN SE AB ALTERO"* =

i. "that which was caused by another to be an actual thing in itself";
ii. "that which was moved by another to stand alone";
iii. "that which receives from another its capacity to continue existing even without being intimately bonded to another.

EXAMPLES: the angels; the human souls; the forms; the generators.

2. *"ENS POTENS ESSE IN SE"* =

i. "that which can, in the future, become a thing in itself";
ii. "a potential being in itself";
iii. "the undeveloped potency of a form or of a generator";[32]

IX. "INDEPENDENCE" =

---

[32] The undeveloped potency of forms or generators is the highest member of the grouping called *"ens in altero"* and the lowest member of the grouping called *"ens in se"*.

i.   "individuality";
ii. "separateness";
iii. "the principle distinction between a non-predicable and a predicable";
iv. "the being of a subject as opposed to the being of an accident";
v.   "being in the primary sense";

EXAMPLES: the being of God, the angels, the human souls, the forms, the generators, the potency of a generator or of a form, and lines of force.

A. "PERFECT INDEPENDENCE" =

i.   "independence of the highest degree";
ii. "the being of a non-inhering subject";
iii. "the being of that which stands apart";
iv. "freedom from the need to be intimately bonded to another";
v.   "existence in the strictest sense";
vi. "substantial being in the primary sense";
vii. "the separateness characteristic of a non-inhering subject".

EXAMPLES: (1) **ACTUALLY** perfect independence: the being of God, the angels, the human souls, the forms, and the generators; (2) **POTENTIALLY** perfect independence: the being of the undeveloped potency of a generator or of a form.

1.   "INFINITELY PERFECT INDEPENDENCE" =

i. "the being of a subject disjoined from *all* other subjects whatsoever, whether inhering subjects or non-inhering subjects";

ii. "the freedom enjoyed by a subject devoid of all intimate bonding to another, *different* subject";[33]

iii. "the being of a substantial subject in which no insubstantial subject is inhering";

iv. "the being of a substantial subject in which all potency has been developed into act";

v. "pure actuality".

EXAMPLES: the kind of separateness characteristic of God and any of the other inhabitants of infinity.

2. "FINITELY PERFECT INDEPENDENCE" =

i. "the being of a non-inhering subject disjoined only from all other non-inhering subjects";

ii. "the being of a non-inhering subject in which an insubstantial subject is inhering";

---

[33] **NOTE OF APRIL 8, 2002:** In The Blessed Trinity, The Three Divine Subjects are not so much *intimately bonded* to one another as they are *totally shared* with one another. Where One begins and ends is wholly and entirely the same for The Other Two. As has often been stated, it is simply a mystery beyond our ken.

    iii.   "the being of a substantial subject in which undeveloped potency is inhering";

    iv.   "impure actuality".

EXAMPLES: the kind of separateness characteristic of forms, generators, human souls, and angels not yet elevated to infinity.

B. "IMPERFECT INDEPENDENCE" =

    i.   "independence of a lower degree";
    ii.   "the being of an inhering subject";
    iii.   "the kind of separateness characteristic of the undeveloped potency of a generator or of a form and of lines of force".

X. "OBJECT" =

    i.   "that which is not disjoined within itself but is disjoined either from everything else whatsoever or from everything other than the insubstantial substrata inhering in it";[34]

    ii.   "that the extremities of which are not intimately bonded to the extremities of anything else whatsoever";

---

[34] **NOTE OF APRIL 20, 2002:** If you prefer say: "that the extremities of which are not tightly stuck to anything else, while whatever is truly inside of its extremities is tightly stuck to something else inside of those extremities".

   iii.   "that the extremities of which are in no way included in, or one with, the extremities of anything else not inhering in it";

   iv.   "a non-inhering non-predicable dissevered at
least from all other non-inhering non-predicables".[35]

A. "INFINITE OBJECT" =

   i.   "that which is without disjunction in itself
and is disjoined from (*i.e.:* not intimately
bonded to) all other subjects *whatsoever*";

---

[35] Only God, angels, human souls, forms, and generators can
legitimately be called "objects". To speak of visible material
subjects (such as individual horses and men) as three-dimensional "objects" is to speak in a very strained manner, since,
from the standpoint of their *spatial* substrata, the extremities
of every three-dimensional subject are joined to, and present
in, the extremities of the non-spatial substratum of a substantial subject. The long sought "ultimate building bricks" of the
universe are *non-spatial* ones (and rather than being the
"building bricks" they are, more correctly, "spiders spewing
out the gossamer threads" which science has long ignored but
which, nevertheless, are truly the insubstantial "stuff" of
which three-dimensional things are made). Being non-spatial,
these "ultimate, irreducible constituents" cannot possibly be
"present in" any three-dimensional subjects—not even as
parts of the whole. Each of them is, rather, united, by means
of its substratum, to the lines of force serving as the substratum of some spatial subject. The explanation of exactly how
extended, insubstantial substrata unite with non-extended,
substantial substrata (*i.e.:* exactly how these two radically
different dimensions of reality connect) is perhaps the supreme achievement of the system set forth here.

ii.   "that the extremities of which are in no way included in, or one with, the extremities of anything else whatsoever";
iii.   "a non-inhering non-predicable dissevered from all other non-predicables".

EXAMPLES: God and any angel or human soul in infinity.

B. "FINITE OBJECT" =

i.   "that which is without disjunction in itself and is disjoined from (*i.e.:* not intimately bonded to) all other *non*-inhering subjects";
ii.   "that the extremities of which are in no way included in or one with the extremities of any other non-inhering subject";
iii. "a non-inhering non-predicable dissevered from (*i.e.:* without intimate bonding to) all other non-inhering non-predicables".

EXAMPLES: any angel, human soul, form, or generator not in infinity.

1. "DIUNE FINITE OBJECT" =

i.   "a non-inhering subject to which is united (*i.e.:* intimately bonded) an inhering undeveloped potency";
ii.   "that which remains two subjects eventhough the extremities of each subject are the same".

EXAMPLES: any angel, human soul, or form not in infinity.

2. "TRIUNE FINITE OBJECT" =

i. "a non-inhering subject to which is united an inhering undeveloped potency and a line of force";

ii. "that which remains three subjects eventhough the extremities of each subject are the same".

EXAMPLES : any generator.

XI. THE HAASIAN BREAK WITH ARISTOTLE:

It can be described in any one of the following nine ways:

A. For Aristotle, all substrata are equally independent and separate. I say they are not.

B. For Aristotle, all substrata are *non*-inhering. I, however, say some substrata must always inhere in a more independent kind of substrata.

C. For Aristotle, *no* substratum is *ever* "present in a subject", which is to say that, for Aristotle, *no* substratum is *ever* incapable of existing apart from intimate bonding to another subject (see quotes iii & iv from Aristotle). I, however, say that some substrata *are* "present in a subject", which is to say that some substrata *are always* incapable of existing apart from another subject (*i.e.:* are unable to

206

avoid annihilation, unless they are intimately bonded to that other subject).

D. For Aristotle, there is *no* twofold division of substrata into: (1) those substrata so *perfectly* independent they *never* inhere in other substrata; and (2) those substrata so *imperfectly* independent that they *always* inhere in the first kind. For myself, there *is* such a division.

E. For Aristotle, there is *no* twofold division of separateness and independence into: (1) **PERFECT INDEPENDENCE**: the kind of separateness characteristic of substrata which do *not* inhere in other substrata; and (2) **IMPERFECT INDEPENDENCE**: the kind of separateness characteristic of substrata which *do* inhere in other substrata. For myself, there *is* such a division.

F. Though Aristotle obviously did not use Latin terms, we can rightly say that, for the Aristotelian system:

   (1)   Every substratum is an *"ens in se"* (*i.e.:* "thing present in itself" as opposed to a "thing present in another");

   (2)   Every *"ens in se"* is a substratum;

   (3)   Every accident is an *"ens in altero"* (*i.e.:* "thing present in another", which is the same as saying "thing present in a subject");

   (4)   Every *"ens in altero"* is an accident.

I agree with #2 and #3 above. I, too, say that every *"ens in se"* is a substratum, and that every accident is an *"ens in altero"*. I deny, however, that every substratum is an *"ens in se"*, and I deny that every *"ens in altero"* is an accident. I say that *some* substrata are *"ens in altero"*, and I say that *some* *"ens in altero"* are substrata.

If an *"ens in altero"* <u>is</u> predicable of that other in which it inheres, it is an accident. If it is <u>not</u> predicable of that other, it is an insubstantial substratum, which is to say it is either a line of force inhering in its generator or the undeveloped potency of a form inhering in its form or the undeveloped potency of a generator inhering in its generator.

To say a substratum is an *"ens in altero"* is to say it *inheres* in that other, which is the same as saying it is "being incapable of existence apart from the said subject". That does not by any means say it is being predicable of that other. [36]

As is manifest, no such distinctions were ever made in the Aristotelian system.

G. For Aristotle, the following are all equivalent terms, which is to say that each signifies the same, identical thing:

---

[36] **NOTE OF MARCH 30, 2002:** Since some are dense beyond belief, let me repeat it *ad nauseam*: When I speak of a substratum as being "incapable of existence apart from the said subject", I mean it is incapable of existence (*i.e.:* incapable of avoiding annihilation) apart from being *intimately bonded* to the said subject. Rest assured, though, that no matter *how* many times I say it, a multitude of even the most educated and intelligent of individuals will still not get it.

1. "that which underlies and is the subject of everything else;" (see quote vi above from Aristotle);

2. "that which is no longer predicated of anything else;" (see quotes i and ix above);

3. "that which exists as underlying;" (see quote x above);

4. "potential substance;" (see quote x above);

5. "that which is *not* present in a subject;" (see quote iii above);

6. "that which is *not* being incapable of existence apart from another subject;" (see quote iv above);

I agree with Aristotle that quotes 1 thru 4 are equivalent. I agree with him that 5 and 6 are equivalent to each other. I do *not* agree, however, that 5 and 6 are equivalent to 1 thru 4 . Contrary to Aristotle, I say that, *in the case of matter*, "that which underlies and is the subject of everything else" *is most certainly also* "being present in a subject", which is to say matter *is indeed* "being incapable of existence apart from" a more independent kind of subject.

Matter (*i.e.:* lines of force) is **not** present in that more independent subject as something present in "that which *underlies*" On the contrary, matter is present in that subject as something present in "that which *generates*". Notwithstanding that distinction, it still remains true that matter is truly *present in* that more independent subject. It is *present in* it because matter is "being incapable of existence apart from the said subject" (**i.e.:** is incapable of avoiding annihilation unless *intimately bonded* to the said subject). After all, unless a generator generates a line of force extending from the generator's undeveloped potency to its actuality, there is no such thing as either matter or any individual material things. That means, of course, that, in the system set forth by this author, matter *can* be annihilated, and, in fact, it *is* repeatedly being annihilated and replaced by newly generated lines of force.

H. For Aristotle, the following terms all signify the same, identical thing:

1. "primary substance;" (see quote ii above);

2. "the individual man" or "the individual horse;" (see quote iii above);

3. "that which is neither predicable of a subject nor present in a subject;" (see quote iii above);

4. "that which does not imply the presence of something in something else, i.e. in something that underlies it;" (see quote xi above);

5. "that which underlies and is the subject of everything else;" (see quote vi above).

For Aristotle, then, each horse, man, and every other individual, three dimensional thing, implies something which is underlying but which is *not* present in something else. Like Aristotle, I, too, say a horse implies an underlying, ultimate substratum which is not present in "something that *underlies* it". Still, that ultimate substratum is "present in something else", because every individual, three dimensional thing implies matter as the ultimate substratum, and matter cannot exist apart from the generators of matter (*i.e.:* matter cannot exist apart from intimate bonding to the generators of matter). Still, the generators of matter do not *underlie* matter; each generator *generates* matter within itself.

The beginning of the problem is this: For Aristotle, there is only *one* way to be "being present in a subject". That is to say, there is, for Aristotle, only *one* way to be "being incapable of existence apart from the said subject" (*i.e.:* only *one* way to be incapable of existence apart from *intimate bonding* to the said subject), which is the same as saying that, for Aristotle, there is only one way to *inhere* in another. For Aristotle, the only way to inhere in something else is to be present in that which *underlies*. For Aristotle, then, it necessarily follows that the idea of an ultimate substratum being present in another can only mean the presence of something in something which *underlies* it. But, that which is present in something which *underlies* it is, by definition, *not* the *ultimate* substratum. Therefore, for Aristotle, because "present in" can only mean "incapable of existence apart from that which *underlies*", the idea of "an ultimate substratum present in something else" is an inherently self-contradictory idea.

Aristotle, though, is in error. "Present in" means, by his own admission, "being incapable of existence apart from the said subject" (see quote iv above). Therefore, it does not by any means *necessarily* mean "being incapable of existence apart from that which *underlies*," unless, of course, we assume, with Aristotle, that the only way to be incapable of existence apart from intimate bonding to a subject is to be present in that subject as that which *underlies*.

I, however, have shown that there is another way to be incapable of existence apart from intimate bonding to a subject. I have shown that a subject can do something besides *underlie* that which can't exist apart from it. I have shown that matter is lines of force which cannot exist apart from the generators which generate those lines of force. Therefore, there is such a thing as a subject which *generates* without *underlying* that which cannot exist apart from intimate bonding to it. [37]

---

[37] **NOTE OF MARCH 30, 2002:** If you say what's present in another is *always* present in what *underlies*, then you at least effectively say that whatever is "present in another", is always a characteristic of the matter of which a material object is made. For example, "present in another" would signify the roundness and smoothness of a glass ball whose matter is smooth, round glass. In saying that, though, you effectively say that the only way to be "present in another" (*i.e.:* the only way to be incapable of existence apart from intimate bonding to another) is to be what some mass of matter *is being*. Unfortunately, the one thing which science's microscopes have made absolutely clear to us is this: Whatever the *ultimate* constituents of any given, material object might be, not a one of them *is being* anything even so little as *remotely* close to any of the characteristics given to our senses. It is, therefore, insane to think that, when we sense the characteristics of

some physical object (whether the glass ball or what have you), that we are sensing what's present in that object's ultimate constituents as "in that which underlies". What, then, _is_ the relationship between the sensed characteristics and whatever might be the ultimate constituents of material things? Because Aristotle can think only in the terms of "present in that which underlies", he can give you no answer. With him, there is an impenetrable wall between the internal characteristics of our sense images and the one factor with which his definitions _demand_ he connect those characteristics—namely: the internal characteristics of the atom's ultimate constituents. In such a circumstance, universal skepticism is the only alternative. After all, the internal characteristics of our sense images are left dangling in the air as inexplicable mysteries which—in some mind-boggling way—are nothing more than a mistake. Oh, what an amazing state of affairs! What we see, we do not see. What we hear, we do not hear. Everything we observe most intensely, we observe not at all.

Fortunately, Aristotle's wall comes tumbling down the instant you realize this: Though the characteristics we sense are present in _matter_ as that which _underlies_, all matter itself is, in turn, something present in another as that which _generates_ rather than _underlies_ matter. Unfortunately, however simple an answer it is, it's one you cannot even begin to grasp if, like Aristotle, you assume that _whatever_ is "present in another" (_i.e.:_ all that is incapable of existence apart from intimate bonding with another) is invariably present in that other as "that which _underlies_".

If every "A" is present in a "B" as that which _underlies_ "A", then every "A" is present in that which _is being_ "A". But if "A" is the internal characteristics we sense, and "B" is the ultimate constituents of the atom, then there is no way "A" is something which "B" _is being_; and so, the only alternatives are either the insanity of universal skepticism or the sanity which says that "A" is present in "_C_" as that which _underlies_, and it is then "_C_" which is present in "B" as that which does something very much other than _underlie_ "C". Realize that, and it becomes quite clear: All of humanity's incessant and

213

I.   For Aristotle, whatever is "being present in a sub-ject" is necessarily "being present in something which underlies it" (see quote xi above). But, for Aristotle, "that which underlies" means the same as "that of which everything else is predicated, while it is itself not predicated of anything else" (see quotes i, ii, iii, vi and viii above). Therefore, for Aristotle, it is necessarily true that whatever is "being present in a subject" must also be predi-cated of that subject.

For Aristotle, then, there is only one way for anything to be "being present in something else": Whatever is "being present in a subject" is being "that which is *predicated* of that subject". I, how-ever, have shown that lines of force are a kind of substratum which is "being present in something else"; and yet, I have shown that, while they are "being incapable of existence apart from intimate bonding to the said generators", they are most cer-tainly not *predicable* of those generators. After all, though the lines of force can only have being as long as they remain within the very heart of the generators, they are merely a form of debris ex-creted by the generators.

END OF DEFINITIONS

---

frantic concern with the internal characteristics of the ulti-mate constituents of the atom (*i.e.:* all the concern with what ultimate substrata *are being*) is utterly irrelevant to the quest for certitude regarding the internal characteristics of our sense images.

(3)
MY DEFINITION OF "SUBSTANTIAL SUBSTRATUM" IS NOT CIRCULAR:

You wrote as follows:

Furthermore, you need to be more careful about the definitions you offer. The one on p. 251 has the air of circularity. You first define "substantial substratum" using the notion of substratum in the definition, and then say that all substrata are substantial substrata.

In the first place, I did not say that all substrata are substantial substrata. I said that, *for Aristotle*, all substrata are substantial. As I wrote on page 251:

. . . we can now state the Aristotelian position on substrata in the following three propositions:

(1) All substrata is (are) substantial substrata.

Secondly, to attempt to define "*substantial* substratum" is to attempt to define a particular <u>*species*</u> of the <u>*genus*</u> "*substrata*". It is an attempt to distinguish between <u>*in*</u>substantial substrata and *substantial* substrata.

Now then, in contrasting the two species of a particular genus, one *usually* limits oneself to merely *stating* the genus in the definition of the species. One does not *usually* restate the definition of the genus in the definition of one of the genus's species.

For example, when asked to explain what "rational animal" means, one would not *usually* say that it is that kind of living, sensing, self-moving, material object which is also capable of reason, and it contrasts with that kind of living, sensing, self-moving, material object which is *in*capable of reason. On the contrary, we usually say that "rational animal" means that species of the genus "animal" which *is* capable of reason, and it contrasts with that species of the genus "animal" which is *not* capable of reason.

By the same token, "substantial substrata" means that species of the genus "substrata" which is *neither* predicable of, *nor* incapable of existence apart from, another member of that genus, and it contrasts with "*in*-substantial substrata" which is that species of the genus "substrata" which, though it is *not predicable* of another member of that genus, *cannot* exist *apart from* another member of that genus.

Perhaps the problem is this: Perhaps you missed it when I defined "substrata". On pages 244 & 245 of my paper, I quoted from Aristotle as follows:

> Now the substratum is that of which everything else is predicated, while it is itself not predicated of anything else.
> ——*Metaphysics*; Book VII; Chap. 3; 1028b: 35. *Great Books*; Vol. 8: pg. 551.

That is the definition of "substrata" which I adopted, as is very plainly stated on page 254 of my paper. In our attempts to avoid circular definitions, we can, if you wish, substitute that definition for the word "substrata" wherever it occurs. It would, however, be unbe-

lievably awkward, to say the least, if we were to say as
follows:

"**SUBSTANTIAL** substrata" signifies that species
of the genus,

> THAT OF WHICH EVERYTHING ELSE IS PREDI-
> CATED WHILE IT IS ITSELF NOT PREDICATED
> OF ANYTHING ELSE,

which is not only *not* predicable of another

> THAT OF WHICH EVERYTHING ELSE IS PREDI-
> CATED WHILE IT IS ITSELF NOT PREDICATED
> OF ANYTHING ELSE,

but which is also *not* incapable of existence apart
from another

> THAT OF WHICH EVERYTHING ELSE IS PREDI-
> CATED WHILE IT IS ITSELF NOT PREDICATED
> OF ANYTHING ELSE.

It contrasts with "**INSUBSTANTIAL** substrata"
which is that species of the genus,

> THAT OF WHICH EVERYTHING ELSE IS PREDI-
> CATED WHILE IT IS ITSELF NOT PREDICATED
> OF ANYTHING ELSE,

which, while it is *not* predicable of another

> THAT OF WHICH EVERYTHING ELSE IS PREDI-
> CATED WHILE IT IS ITSELF NOT PREDICATED
> OF ANYTHING ELSE,

is, however, incapable of existence apart from a substantial substratum.

Furthermore, on pages 244 and 245 of my paper, I gave several quotes from Aristotle in which the terms, "anything else", "stratum", and "subject" are used interchangeably, as Aristotle refers to both substance and substratum as being that which is neither predicable of, nor present in, another. Therefore, it certainly seems that I would be doing nothing but following in Aristotle's footsteps were I to say that "substantial substratum" means that which is neither predicable of, nor incapable of existence apart from, any other *thing, subject*, or *stratum*.

The problem here is this: For Aristotle, as for every other human being *save myself*, "substrata", as a *general* term, is *sui generis*: It does not admit of species. For Aristotle, as for every other human being *save myself*, "substrata" and "that which is *neither* predicable of, *nor* incapable of existence apart from, a subject" are *absolutely equivalent* terms. For Aristotle, as for every other human being *save myself*, there can be no such thing as a kind of substratum which is <u>in</u>capable of existence apart from a subject of which it can *not* be *predicated*.

For the sake of clarity, let me attempt to say the above in another fashion. I say that Aristotle and I agree that every substratum is a "non-predicable", if I may use such a term. In other words, for both Aristotle and myself, every substratum is strictly a *subject* of predication and is never that which is predicated of a subject. I, however, then go on to say that "non-predicables" (*i.e.:* "subjects") can be of more than one kind. On the one hand, one has *"perfectly* independent non-predicables". A "perfectly independent non-predicable" is that kind of "non-predicable" which *can* exist apart from all other "non-predicables". On the other hand, one has *"imperfectly* independent non-predicables". An "imperfectly independent non-predicable" is that kind of "non-predicable" which *cannot* exist apart from *all* other "non-predicables". On the contrary, eventhough it is a "non-predicable" and a "subject", it can only have being by inhering in a "perfectly independent non-predicable". For Aristotle, however, all of that is insane, because, for him, *whatever* is a "non-predicable" is also *necessarily* "that which is *not in*capable of existence apart from a subject". In other words, for Aristotle, *every* "non-predicable" *must* be "that which does *not* inhere in a subject"; whereas, for myself, *only substantial* non-predicables *must* be "that which does *not inhere"*.

This is *my* unique contribution to philosophy. It is my contention that I am the first human being in the history of mankind to introduce the idea of more than one kind of substratum, which is to say more than one kind of "non-predicable", which is the same as saying more than one kind of "subject". I am the first man in history to take the group,

THAT OF WHICH EVERYTHING ELSE IS PREDICATED WHILE IT IS ITSELF NOT PREDICATED OF ANYTHING ELSE

and to introduce a threefold division into it as follows:

SUBSTRATA (NON-PREDICABLES)

A. SUBSTANTIAL SUBSTRATA (*i.e.:* non-predicables which either can be in the future, or currently are, not inhering in another non-predicable)

219

I. **ACTUALLY SUBSTANTIAL SUBSTRATA (*i.e.:* non-predicables which are actually not currently inhering in another non-predicable).

II. **POTENTIALLY SUBSTANTIAL SUBSTRATA (*i.e.:* non-predicables which, though they can become actually substantial substrata, are, here and now, annihilated unless they are inhering in an actually substantial substratum).

B. INSUBSTANTIAL SUBSTRATA (*i.e.:* non-predicables which can never have being save as that which inheres in an actually substantial substratum).

Finally, I am the first man in history to explain how it is possible to have a kind of substratum which is not predicable of the subject in which it must inhere (*i.e.:* to which it must be intimately bonded) in order to have being. For, I have explained that each unit of insubstantial substrata is a line of force generated between the actuality and the potentiality of a unit of substantial substrata when that unit of substantial substrata acts in the presence of a drag inducing differential between its actuality and its potentiality.

(4)
MEANING OF "THING" IN "ONE THING BUT TWO ESSENCES":

You wrote:

What do you mean when you say that a substantial substratum is one thing but has two essences. First, it is not clear what 'thing' means in this context.

In the Aristotelian system, "thing" has two different meanings. On the one hand, there is the "primary, truest, simplest, most definite, and unqualified sense". In that case, "thing" means "that which is *not* predicated of a stratum, but of which all else is predicated". On the other hand, there is the "secondary, limited and qualified sense" of the word "thing". In that case, "thing" means "that which *is* predicated of a stratum (*i.e.:* predicated of a subject)".

It is generally recognized, however, that there is a third meaning possible to the word "thing". In this case, "thing" means "that which is undivided in itself and divided from all others". More exactly, it means "a subject undivided in itself and disjoined from (*i.e.:* without intimate bonding to) all other subjects". When so defined "thing" necessarily means *"perfectly independent* thing", and it is equivalent to such terms as "object" and, at least for some, "hypostasis".

What do we mean, though, when we use the phrases "a subject disjoined from all other subjects" and "a subject without intimate bonding to all other subjects"? Since my system admits of several kinds of subjects, the definition of an object requires further clarification of the term "subjects".

Clearly, in my system, some subjects inhere in other subjects. If, however, one is to have a subject which is undivided in itself and disjoined from <u>all</u> other subjects, one must have a subject which *neither* inheres in any other subjects, *nor* has any other subjects inhering in itself. In my system, such a condition can occur only where all undeveloped potency has been converted into actuality, which is the same as saying that such a condition exists only where substantial substrata are *pure* actuality. In my system, such a condition is found only in infinity.

One can perhaps readily see that, in my way of thinking, objects must be divided into two kinds. **FIRST**, we will have what we can call "infinite objects". An infinite object is any subject undivided in itself and divided from *all* other subjects *whatsoever*, which is the same as saying "a non-predicable dissevered from all other non-predicables". **SECONDLY**, we can have what we shall call "finite objects". A finite object is any *substantial* subject undivided in itself and disjoined *only* from all other *substantial* subjects. In other words, "finite object" is equivalent to "a non-inhering subject disjoined from all other *non-inhering* subjects". It is also equivalent to "a non-inhering non-predicable dissevered from (*i.e.:* without intimate bonding to) all other non-inhering non-predicables".

Obviously "finite object" does not exclude a *non*-inhering subject in which an *inhering* subject is inhering. All that's required is that the non-inhering subject be not joined to another *non*-inhering subject.

For the sake of practicality, let us agree that, whenever we use the term "object", it shall be understood that we mean *"finite* object". If we wish to speak of an *infinite* object, we shall use the term "infinite object".

Turn now to Aristotle. As I've said repeatedly, in the Aristotelian system, the genus "non-predicables" (*i.e.:* "subjects") admits of no species. As a result, in the Aristotelian system, every "non-predicable" (*i.e.:* every "subject") is necessarily "that which is *neither* predicable of, *nor* inhering in, another subject", which is the same as saying that, for Aristotle, *every* subject is *necessarily* disjoined from *all* other subjects *whatsoever*. Of necessity, that means that, for Aristotle, *whatever* is a "thing in the primary sense" is also necessarily a subject disjoined from (*i.e.:* without intimate bonding to) *all* other subjects *whatsoever*. In effect, whether he realized it or not, Aristotle was saying:

(1)   the only things which are things in the primary sense of the word "thing" are the inhabitants of infinity;

(2)   God is the only object;

(3)   nothing is independent save that which partakes of the divine nature.

Those are the *necessary* conclusions of saying that every "thing in the primary sense" is a subject disjoined from *all* other subjects. That is to say, they are the necessary conclusions of saying that every "thing in the primary sense" is a subject not *intimately bonded* to any other subject whatsoever. Had Aristotle never seen Siamese twins?

Let Aristotle say what he will. I say that it is by no means true that *every* "thing in the primary sense" is also a "perfectly independent thing", whether "perfectly independent thing" means "subject disjoined from all other subjects *whatsoever*", or whether it means "non-inhering subject disjoined from all other non-inhering subjects". On the contrary, while it is true to say that every "perfectly independent thing" is also a "thing in the primary sense", the assertion is not convertible; and so, some of those things which are "thing in the primary sense" are not things in the sense of "perfectly independent thing", which is the same as saying not every subject is either a subject disjoined from all other subjects or a non-inhering subject disjoined from all other non-inhering subjects. After all, some subjects are inhering subjects intimately bonded to non-inhering subjects.

Now then, when I say that every *substantial* substratum is *one* thing, I mean it is one *perfectly independent* thing in the sense that it is one non-inhering subject disjoined from all other non-inhering subjects. In other words, I mean it is one *object*. **But**, this *one independent object* includes within itself three things each of which is a "thing in the primary sense". Since "thing in the primary sense" means a "subject of predication" as opposed to "that which is predicated of a subject", we now have three subjects joined together in one perfectly independent object. That does not mean, however, that we have *three subjects* joined together to form *one subject*. By no means! What we have is three subjects joined together in one independent *object*; **But** that *object* is not itself a *subject*. At least, it is not a subject of which the other subjects can be predicated *in any way*. By that, I mean that our subjects are neither parts joined together to form the one whole object nor characteristics of the one whole object.

It might simplify the above to say this: What we have here is more akin to one perfectly independent subject carrying on its back two parasites (*i.e.:* we have a subject which is "perfectly independent" and "substantial" because it is *disjoined from* all other *non-inhering* subjects). Like all parasites, each of these two parasites is another subject which cannot be, and must not be, confused with the host to whom they are most intimately united. Unlike all other parasites, however, these two parasites are *so* intimately united, that their very being must constantly be fed to them by their host. One of the parasites *never will be able* to have being in any other fashion save by inhering in the perfectly independent subject while one end of itself is firmly attached to the other parasite. The other parasite may one day become a perfectly independent subject itself, but, for the moment, it has no being except in so far as it inheres in the perfectly independent subject.

(5)
HOW INDIVIDUATE 1 THING IF IT HAS 2 ESSENCES:

You wrote:

> . . . you do not explain how one thing can be individuated, when it has two essences. What role does the essence play, if it does not give the ground for individuating a separate thing?

The answer to this problem lies within the answer we have just given to the previous problem. Let us attempt to unravel it.[38]

---

[38] **NOTE OF MARCH 31, EASTER SUNDAY, 2002:** These days, the answer which follows above strikes me as being far less clear and effective than it should be. I now think it would have been better had it gone something like this: I never said one *thing* "can be individuated, when it has two essences" (Mr. Scaltsas' words above). At least, I never *intended* to say such. What I at least *intended* to say is this: An *object* can be individuated despite the fact that it may enclose within its extremities more than one *subject*. To illustrate what that means, imagine two Siamese twins, George and John, so

225

completely joined together that, though each has his own head, they have but one torso, one set of arms, one set of legs, and one set of internal organs. In such a case, what "individuates" George from John is very different from what individuates their common body from all other human bodies. George and John are each a *subject*, and their common, human body is an *object*. What individuates their body from all other visible human bodies is the fact that their body is not *intimately bonded* to any other visible body whether human or otherwise. Conversely, what individuates George and John from one another is that each has his own very distinct and very different essence—a fact manifested by the way they disagree with each other on many issues. Essence is what *always* individuates *subjects* from one another regardless of whether or not they are intimately bonded to one another. In other words, to use Mr. Scaltsas' above words, essence *does* always "give the ground for individuating a separate thing", *provided* "separate thing" means a thing which is separate in the sense that its *identity* is different from all other identities. However, essence is *not always* what individuates *objects* from one another. In other words, to use Mr. Scaltsas' above words, essence does *not* always "give the ground for individuating a separate thing", **_if_** "separate thing" means a thing which is separate in the sense that its extremities are not intimately bonded to any other set of extremities. That's because, within the boundaries established by their extremities, some *objects* (**ex gr.:** Siamese twins) enclose more than one *subject*. Naturally, if a given object encases only a *single* subject, then it has an essence, and its essence is both that which makes it a *subject* and that which individuates it from all other subjects. Still, its essence is not what makes it an *object*. What makes it an *object* is a set of extremities *not intimately bonded* to any other set of extremities. On the other hand, if a given object encases *multiple* subjects, then, *strictly* speaking, that object *itself* does not <u>have</u> an *essence*. It merely <u>encases</u> two or more essences, and those encased essences individuate the encased *subjects* from one another. Those encased essences have nothing to do with individuating the *object* itself from all other *objects*. Always and everywhere, what individuates one *object* from all other *objects* is a set of extremities **NOT INTI-MATELY BONDED** to any other set of extremities. There-fore, in the case of <u>objects</u>, "thing" means: "that the extremi-ties of which are not intimately bonded to any other extremi-ties; whereas, whatever is encased within its extremities is intimately bonded to something else within those extremi-ties". In the case of <u>subjects</u>, "thing" means: "that which is so individuated by its essence, that it is not *predicable* of an-other." At least, that's the way it is for <u>me</u>. For Aristotle and those who think in his exceedingly crippled manner, "thing"—taken in its strictest sense—can only mean: "that which is so individuated by its essence, that it is **NEITHER PREDICABLE** of another **NOR** incapable of existence apart from intimate bonding to another." It is no wonder, then, that, when confronted with "Haasian" types saying that—even when taken in its strictest sense—"thing" applies to *subjects* one way and to *objects* another, the Aristotelian types are utterly at a loss to make any sense of what is being said. They seem to suffer some kind of *irremovable* mental block which makes it impossible for them to perceive that—taken in its strictest sense—"thing" *necessarily* implies *only* that which is *not predicable* of another. But, among non-*predica-bles*, some are *additionally* non-*inhering*, which is to say that they are **NEITHER PREDICABLE** of another **NOR INCA-PABLE OF EXISTENCE APART FROM INTIMATE BONDING TO ANOTHER.** Since, then, *not all* non-*predica-bles* are also non-*inhering*, it necessarily follows that—*even when taken in its strictest, truest, primary, most definite and unqualified sense!*—the word "thing" has two radically differ-ent meanings, just as the word "animal"—*even when taken in its strictest, truest, primary, most definite and unqualified sense!*—has one meaning when applied to dogs and a very dif-ferent one when applied to humans. For "Haasian" types, it is a distinction as clear and easy to grasp as distinctions ever get to be. For Aristotelian types, though, it is a distinction they cannot grasp no matter how plain or numerous the ways it is stated. How is that possible, unless it is a distinction they

SUBSTRATA

If the word "essence" be taken in the primary sense of the word "essence" (*i.e.:* if it be taken to apply to subjects rather than to accidents), then I most heartily agree with you that the essence does indeed, *always*, as you say, "give the ground for individuating a separate thing". The problem, however, is this: When you and I speak of individuating a **SEPARATE THING**, our words are exceedingly far from having the same meaning.

I say a thing can be "separate", "independent", and "individual" in three radically different ways. Let us examine those three ways.

In the first place, some things are what we shall call "*imperfectly* separate, independent, and individual". By that, I mean they are separate and independent individuals in the sense that they are subjects of accidents and subjects of predication as opposed to accidents and predicates of a subject. That's the same as saying they are separate, independent and individual in the sense that they are things in the primary sense of the word "thing". They are by no means separate, independent and individual in the sense that they are *not intimately bonded* to another. Only things in the secondary sense of the word "thing" (*i.e.:* only accidents) are *absolutely* *non*-separate, *non*-independent *non*-individuals, and that's because only they are *both* intimately bonded to another *and* predicable of that other to which they are intimately bonded.

In the second place, some things are what we shall call "*perfectly* separate, independent, and individ-

---

will not see? In the final analysis, what their stance strongly *suggests* is a level of intellectual deadness and downright perversity so pronounced it defies both the capacity of the human *mind* to *comprehend* it and the capacity of the human *tongue* to *describe* it. God have mercy on them, if that is the case!

ual *but* on the *finite* level". By that, I mean they are *non*-inhering subjects disjoined from (*i.e.:* without intimate bonding to) all other *non*-inhering subjects. Such subjects, of course, are free to be intimately joined to the subjects *inhering* in them.

Finally, some things are what we shall call "*perfectly* separate, independent, and individual *but* on the *infinite* level". By that, I mean that they are non-inhering subjects disjoined from *all* subjects *whatsoever*, whether inhering or non-inhering subjects (*i.e.:* they are subjects not *intimately bonded* to any other subject whatsoever).

Clearly, only that last classification is independent, separate, and individual in the most exactingly strict application possible to those terms. Still, the other two classifications remain independent, separate and individual in a very legitimate sense, since there is still an immense difference between the status of the accidents of a subject and a subject of accidents. That being the case, only an *unreasonably* literal person would, in my opinion, insist that only infinite independence is legitimately called independence.

Perhaps you now begin to see the difference between my position and your Aristotelian one. My position is this: If "separate thing" and "individual thing" be taken to be equivalent to "thing in the primary sense" and to "subject of accidents as opposed to accidents of a subject", then I must wholeheartedly agree with you that essence *always* does "give the ground for individuating a separate thing". After all, in such a situation, essence can individuate any one of four general kinds of subjects (*viz.:* a group of lines of force disjoined from all other such groups; a unit of undeveloped potency disjoined from all other such units; a substantial subject disjoined from all other *substantial* subjects; and a substantial subject disjoined from all other subjects *whatsoever*). Oppositely, if "separate thing" and "independent thing" be taken to be equivalent to "subject disjoined from all other subjects whatsoever", then I *must* deny that essence always gives the ground for individuating a separate thing, since, in my way of thinking, an essence sometimes gives the ground for individuating one subject from a second subject so intimately conjoined to it, that they cannot be called more than one object.

Contrary to what Aristotle has caused the world to think, *material* subjects (such as "this horse" and "this man") are not as independent as God is. Individual material things are independent things only to a very limited degree. They are independent only in the sense that their substratum does not inhere in another substratum *of the same general species.* "This horse" is independent only in the sense that its substratum does not inhere in (*i.e.:* is not intimately bonded to) the substratum of another horse or in the substratum of the man on its back or in the substratum of the ground beneath its feet or in the substratum of any of the other spatially material subjects which our senses can immediately experience. Nevertheless, the substratum of the horse is many lines of force which will cease to be, unless an astronomical number of sub-atomic particles maintain a certain proximity to one another and, thereby, generate those lines of force between them.

Contrary to what most men think, the substratum of our bodies is not the minute particles which come into view when electron microscopes are focused upon our bodies. Those suddenly visible molecules, atoms, and sub-atomic particles are not subjects present in our bodies as parts capable of being divided from their whole. On the contrary, they are that in which the matter of man's body inheres.

For you, as for your ancient mentor, every "thing in the primary sense" must be a subject disjoined from all other subjects whatsoever, which is to say a subject not *intimately bonded* to any other subject whatsoever. Like Aristotle, whenever you think of a thing in the primary sense, you always assume that your subject must be an independent thing in the most narrow sense of the word "independent" that your mind can possibly conceive. You cannot imagine a subject which, while it is still strictly a *subject* of prediction, is still so imperfectly independent that it cannot have being unless it inheres in another kind of subject which, though it is more independent than its inhering subject, is, still, not so independent as to be disjoined from *all* other subjects.

For that reason, like Aristotle, you also assume that "essence in the primary sense" can only individuate a subject disjoined from *all* other subjects. ***But***, you are both mistaken. Parasites, too, are "things in the primary sense". Consequently, "essence in the primary sense" can indeed individuate a thing which, though its matter cannot have being save it inhere in a radically different kind of substratum, is, nevertheless, a "separate" and "independent" thing in the *imperfect* but still *legitimate* sense that it is a subject of accidents rather than an accident of a subject.

(6)
NEW TERMS NEED EXAMPLES, ESPECIALLY "DIUNE" & "UNIFORM":

You wrote:

> I think it would help the reader very
> much if you spent more space in giving
> examples of what you mean. Particularly
> when you introduce new terms to the lan-
> guage such as 'diune'; but also 'uniform'
> (pp. 256 & 257), etc.

To begin with, I nowhere stated that substratum
is "diune but also uniform" (assuming that's what you
meant). On page 251, I said that, *for Aristotle*, all substrata are absolutely uniform.[39] As for my own position
on this issue, I said, on pages 256 & 257 of my paper,
that every indivisible unit of finite substantial substrata
is always *at least* diune *rather than* uniform.

If you look carefully, you will see that this, too, is
but another application of the problem we have been
discussing above. To speak of an "indivisible unit of finite substantial substrata" is to speak of an *object* (*i.e.:
finite* object). In other words, we are speaking of a substantial substratum considered as "that which is disjoined from all other *substantial substrata*". In that case,
we are talking about an object *describable* as, but *not
composed* of, a non-inhering subject in which there is
an inhering subject. When we speak in that fashion, the
object is indeed "one thing" which is intrinsically "di-
une". It is such because the two subjects comprising
that one object are—notwithstanding their conjunction
in one object—still two *separate* subjects, due to the fact
that they are *separate* in *identity*. As I have repeatedly

---

[39] **NOTE OF APRIL 8, 2002:** In saying all substrata are absolutely uniform for Aristotle, I'm saying that, for him, every
substratum is always of the same kind—namely: that which is
neither predicable of another nor incapable of avoiding annihilation unless it coheres to another.

232

pointed out, they are *separate* because neither is either a predicable of the other or a predicable of the whole. To be <u>not</u> *intimately bonded* to another is one kind of *separate* thing; to be intimately bonded to another without being predicable of that other—that is a very different but very legitimate kind of *separate* thing.

In short, the distinction between "diunity" and "uniformity" is the same as the distinction between:

(1) a philosophical system which says every object is of necessity a subject dissevered from all other subjects, which is to say a subject not intimately bonded to any other subject;

AND

(2) a philosophical system which says that every *finite* object is, of necessity, at least two subjects, one of which—though it inheres in the other (***i.e.:*** is so intimately bonded to the other, it will be annihilated if the bond is broken)—is—*notwithstanding that fact!*—still very much separate from, and independent of, the subject in which it inheres. Because it is intimately bonded to the subject in which it inheres, it is not *actually* and *physically* separate from, or independent of, the subject in which it inhere. Nevertheless, since it is separate in identity and has an essence which can in no way be predicated of the subject in which it inheres, it is *logically* separate from, and independent of, the subject in which it inheres.

(7)
2 MEN FUSED TOGETHER &
A PIECE OF EXCREMENT:

Let me now attempt to summarize all of the above in the following, final re-statement.

For Aristotle, there can be no such thing as a "subject of accidents" which is *not also* an infinite *object* (*i.e.:* a subject the extremities of which are not intimately bonded to the extremities of any different subject *whatsoever*). For Aristotle, there can be no such thing as an individual substance (**ex. gr.:** "this man" or "this horse") which—*from the standpoint of its substratum!*—cannot have being unless it inheres in the substratum of another individual substance.

Imagine that!!! An individual substance which—*from the standpoint of its substratum!*—is so present in the substratum of another individual substance, that its substratum will cease to subsist, **if** its substratum ever ceases to inhere in the substratum of that other individual substance!!! For all of that, the inhering substratum of the first individual substance cannot be in any way identified with the substratum of that second individual substance in whose substratum the first individual substance's substratum is inhering. Though the two substrata of the two individual substances are so inextricably united as to allow us to call them one *object*, nevertheless, they are quite distinct one from the other; they remain so radically different one from the other as to defy the imagination; and each continues to serve only one of two very different masters.

It is *somewhat* as though one were to take a white man and a black man, to stand them back to back, and to fuse them together in such a way that there remained only one object which, on one side, remained the front side of a white man looking in one direction, while, on the opposite side, it remained the front side of a black man looking in the opposite direction. There is still only one body having only two arms, two legs, one head and one torso. But, there are still two faces, two individuals, two personalities, two wills, two intellects, two persons, and two colors. The only other thing to bear in mind is this: If we try to separate the two men from one another in such a way as to make them two disjoined bodies, one of the two men will instantly *and necessarily* be annihilated, which is to say he will lapse into nothingness without the tiniest trace. The other will continue to exist, but will do so as a very crippled, and incomplete individual who can never experience any kind of change whatsoever.

At the same time, it is possible that one of the two men will eventually absorb the other man wholly and entirely into himself. When that occurs, there will be only one man in that one object, and that "one man in one object" will of necessity be residing in infinity forever.

Finally (and this is the more difficult part of the concept), as these "two finite men in one object" progress toward the status of "one infinite man in one object," their actions precipitate between themselves (but still within the confines of the one body) a kind of refuse or excrement which is in no way whatsoever either a part of, or a characteristic of, either one of them. In that *qualified* sense, it is a separate and independent subject, eventhough, it would lapse into nothingness without the tiniest trace, if it were—for however brief a moment—separated from either both or any one of its two hosts.

This "third subject within a single object" is what can serve as a "piece of a third man". We must now bring together an astronomical number of these "two finite men in one object", together with their "pieces of excrement", and hold them together within a certain proximity to one another. That is the function of a "form" (a topic not discussed in any real detail in my paper). But, once that astronomical number of "triune objects" (triune because each object is two men and their piece of excrement) is brought together, their now astronomical number of pieces of excrement will constitute the body of a third man who, unlike each of the two men in one object, is *not* inextricably united "back to back" with one opposite type of man. For all of that, from the standpoint of each of the many pieces of matter constituting the substrata of his body, that third man is a single man so inextricably and intimately joined to each of an astronomical number of units of "man & counterman", that his body cannot possibly avoid annihilation, unless his substrata ever continue to inhere in those other units of "man & counterman".

This, then is the idea: Each and every piece of matter always comes in a package _describable_ as, *but not _composed_ of*, three different individuals each having its own set of internal characteristics radically different from the other two, but, at the same time, being so inextricably united to each other, that it is impossible to pull them apart from one another.

Such a concept of three substrata, three subjects, and three individuals in one object is no more possible either to Aristotle or to his philosophical system than it is to the man in the moon. Furthermore, the inability of the Aristotelian system to allow for such a concept makes it impossible for any modern man—trapped within the confines of the Aristotelian system—to understand the true relationship between the individual substances we experience daily and the sub-atomic particles which *seem* to reside within them. Indeed, it misdirects his understanding of that relationship in such a way, as to make it inevitable that—if the individual is truthful with himself about the implications of the Aristotelian system—then he cannot possibly escape the lethal grip of universal skepticism.

I ought to point out in passing, at this point, that the Catholic doctrine of the Blessed Trinity says God is, in effect, three individuals in one object. But, each one of those individuals has the same, one, identical set of internal characteristics. That is very different from what I'm saying about matter. I'm saying matter is *comparable* to the idea of two individuals and a piece of a third all in one object, and each of those three has a set of internal characteristics which is radically, radically different from the other two sets. Nevertheless, like the Blessed Trinity, each of those three subjects in one object is a very distinct individual which cannot be confused with, or in any way predicated of, the other two. Again, the Blessed Trinity is, so to speak, identical triplets conjoined in one substratum in one object. Wherever there is matter, though, there are three utterly disparate substrata being three utterly disparate "individuals" in one object. In short, rather than being a set of identical triplets, they are as different from one another as are a man, a dog, and a pancake.[40]

---

[40] **NOTE OF APRIL 29, 2002:** In the last few days, it has suddenly come to my attention that I have no passage in which Aristotle *specifically* associates the phrase "that which is neither predicable of a subject nor present in a subject" with either one of the two Greek terms generally translated as "substratum" (*viz.: hupostasis* and *hupokeimenon)*. I have only 2 in which he associates the phrase with the Greek term generally translated as "substance" (*viz.: ousia*). Both are from

(8)
THE BITTER TRUTH OF REJECTION:

You wrote:

Unless you can make the text readily readable by the reader, people will not take the time to work through your theory. This is a bitter truth we have all found out about the hard way.

Categories Chap. 5, with one at 2a: 11 and the other at 3a: 6 (2a: 11 and 3a: 6 are the way August I. Bekker, 1785-1871, numbered the pages and lines of Aristotle's works.). However, Aristotle *clearly implies* the connection. For, looking at the Greek text (It's available at www.perseus.tufts.edu on the internet.) for Chap. 9 of Book V of **Metaphysics** (1017b: 24), certain it is that Aristotle associates the Greek word "*hupokeimenon*" with the phrase "which is no longer predicated of anything else". Again, in the Greek text for Chap. 3 of book VII (1028b: 35) he associates the Greek word "*hupokeimenon*" with the phrase "that of which everything else is predicated, while it is itself not predicated of anything else", and then, a mere 7 lines later, at 1029a: 7, says almost the same of "*ousia*"—namely: "that which is not predicated of a stratum but of which all else is predicated". Furthermore, at 1017b: 24, "*hupokeimenon*" is the "ultimate" (*i.e.:* "*eschato*") substratum (*i.e.:* underlayment). Since, for Aristotle, "present in" means present in that which *underlies* (1037b: 2), it is, for him, self-contradictory to speak of an *ultimate underlayment* as present in something else *underlying*. It is, therefore, conclusive that, for Aristotle, both "*ousia*" and "*hupokeimenon*" mean "that which is neither present in nor predicable of another".

My dear Mr. Scaltsas, it is by no means a "bitter" truth to a man of my forty-seven years. When I was a young man, I was a nobody who had never achieved anything. Consequently, I could not stand to look at myself in the mirror, because what I saw always looked far too worthless to me. In my madness, I ignorantly imagined that, if I could but achieve fame as a great thinker, such would give me the power to look at myself in the mirror with pride and satisfaction. Understandably, then, I was frantic to have others "work through my theory", and it was, indeed, a bitter pill to swallow when fifteen years of effort met with nothing but rejection and assurances that I was nothing but an ignoramus revealing the extent of his ignorance.

In 1975, when I was thirty-nine years of age, my late father went insane. I who had hated my parents all my life put aside that hatred and decided to provide my father with the assistance he obviously needed most desperately. I preserved and enhanced his moderately large estate till the day he died. With him gone, I now see to it that, as he requested, his wife, my mother, shall also be well taken care of till the day she dies.

In the act of repaying my parents' concern for my physical well-being as a child, I have found all I *really* need to find, and I have accomplished all that I *really* need to accomplish. All the rest is anti-climactic.

These days, I write because every man owes it to the world around him not to hide his thoughts beneath a basket. I owe it to the world around me to make a *reasonably diligent* effort to make my thoughts available to my fellow man. That, however, is *all* I owe. It is not incumbent upon me either to *force* the world into hearing and understanding me or to ruin my physical and mental health in an *attempt* to compel them to listen.

My thoughts *are* in the open. They have been *reasonably* well stated and are well on their way to being *reasonably* well disseminated. It is for the world to do the rest. I do not even attempt to judge for myself whether or not my thoughts are worthy of any attention. It is merely my duty to throw them abroad where others will easily have the opportunity to do with them whatever *they* want to do with them. If, therefore, the world chooses to ignore what I've written, then so be it. Maybe indifference is all my work deserves. Who cares? All that's important is that I did what charity urged me to do.

On the other hand, if the world's decision to ignore me is a rash and erroneous one . . . . Again! Who cares? I have done my part. It is the world which will have to answer for it if it shall have ignored what could have helped it.

For the rest, I am far more abundantly blessed than the vast, vast majority of my contemporaries can ever hope to be blessed. I have excellent health, an extremely active mind, property, financial independence, high respect and regard in the eyes of both family, friends, and self, and I have a marvelous sense of being at peace with God. Finally, I have reached that sublime state in life when the wisdom of years finally teaches the vainglorious fool that anonymity is the only reliable guardian of a life of unbelievable tranquility. In short, dear sir, no matter how the world reacts to my thoughts, I cannot possibly lose or gain anything which *truly* matters.

Cordially and gratefully yours,

EDWARD N. HAAS

Part 5:

ARISTOTLE'S FUNDAMENTAL FALLACY

(A)

At issue here is one's basic ideas regarding substance *__as substrata__*. Unfortunately, Aristotle, in his work **Metaphysics** presents his ideas on this issue in a manner which leaves much to be desired. It consequently remains to be seen whether anyone can definitively state much about his basic ideas in this area. Nevertheless, let us at least attempt to draw a clear picture of one interpretation of Aristotle's notions about substance as substrata.

Imagine we are conversing with Aristotle. We hold up a solid, stainless steel ball which is three inches in diameter. We ask: "What is this?"

It is my contention that Aristotle would reply: "Taken in its simplest sense, that is a substance being spherical in shape, being silvery in color, being three inches in diameter, and being at rest in your hand."

That little dialogue seeks, first of all, to illustrate that, for Aristotle, every *thing* is—in the simplest sense—what substance is being, and whatever else is—in some other sense—is merely a property of substance. The following quotes from Aristotle's **Metaphysics** at least lend strong support to the contention that such is his position.

> Now there are several senses in which a thing is said to be first; yet substance is first in every sense—(1) in definition, (2) in order of knowledge, (3) in time. For (3) of the other categories none can exist independently, but only substance. And (1) in definition also this is first; for in the definition of each term the definition of its substance must be present. And (2) we think we know each thing most fully, when we know what it is, e.g. what man is or what fire is, rather than when we know its quality, its quantity, or its place; since we know each of these predicates also, only when we know *what* the quantity or the quality *is*.
>
> And indeed the question which was raised of old and is raised now and always, and is always the subject of doubt, viz. what being is, is just the question, what is substance?
>
> ——**Metaphysics**; Book VII; Chap. 1; 1028a: 31-38 & 1028b: 1-4. **Great Books Of The Western World**; Vol. 8; pg. 550

lower right; Encyclopedia Britannica Inc.;
Chicago, 1952.

Clearly, then, only substance is definable. For if the other categories also
are definable, it must be by addition of a
determinant, e.g. the qualitative is defined
thus, and so is the odd, for it cannot be
defined apart from number; nor can fe-
male be defined apart from animal.
——*Metaphysics*; Book VII; Chap. 5;
1031a: 1-3. ***Great Books***; Vol. 8; pg. 553
lower right.

For as 'is' belongs to all things, not however in the same sense, but to one sort of
thing primarily and to others in a secondary way, so too 'what a thing is' belongs
in the simple sense to substance, but in a
limited sense to the other categories. For
even of a quality we might ask what it is,
so that quality also is a 'what a thing is',—
not in the simple sense, however, but just
as, in the case of that which is not, some
say, emphasizing the linguistic form, that
that which is not *is*—not *is* simply, but *is*
non-existent; so too with quality.

We must no doubt inquire how we
should express ourselves on each point,
but certainly not more than how the facts
actually stand. And so now also, since it is
evident what language we use, essence
will belong, just as 'what a thing is' does,
primarily and in the simple sense to substance, and in a secondary way to the

other categories also,—not essence in the simple sense, but the essence of a quality or of a quantity.

——*Metaphysics*; Book VII; Chap. 4; 1030a: 21-30. *Great Books*; Vol. 8; pgs. 552 bottom right & 553 top left.

. . . this is evident, that definition and essence in the primary and simple sense belong to substances. Still they belong to other things as well, only not in the primary sense.

——*Metaphysics*; Book VII; Chap. 4; 1030b: 5-6. *Great Books*; Vol. 8; pg. 553 lower left.

Clearly, then, definition is the formula of the essence, and essence belongs to substances either alone or chiefly and primarily and in the unqualified sense.

——*Metaphysics*; Book VII; Chap. 5; 1031a: 11. *Great Books*; pg. 554 top left.

Aristotle at least sometimes defines "substance" and "substratum" the same way. Of each he says the difference between it and "everything else" is that it is not predicable of a subject (*i.e.:* stratum), while everything else _is_ so predicable. The next two quotes confirm that.

The word 'substance' is applied, if not in more senses, still at least to four main objects; for both the essence and the universal and the genus are thought to be the substance of each thing, and fourthly the substratum. Now the substratum is

that of which everything else is predicated, while it is itself not predicated of anything else. And so we must first determine the nature of this; for that which underlies a thing primarily is thought to be in the truest sense its substance.
——*Metaphysics*; Book VII; Chap. 3; 1028b: 33-1029a: 1. **Great Books**; Vol. 8; pg. 551 bottom left and top right.

We have now outlined the nature of substance, showing that it is that which is not predicated of a stratum, but of which all else is predicated.
——*Metaphysics*; Book VII; Chap. 3; 1029a: 7. **Great Books**; Vol. 8; pg. 551 top right.

Turning momentarily to Aristotle's **Categories**, a part of his larger work, the **Organon (*i.e.: Logic*)**, we learn that substance has a second critical distinction. Not only is substance that which is not predicable of a subject (***i.e.:*** stratum); but rather, it is also that which is not present in a subject.

Substance, in the truest and primary and most definite sense of the word, is that which is neither predicable of a subject nor present in a subject; for instance, the individual man or horse.
——*Categories*; Chap. 5; 2a: 11-13. **Great Books**; Vol. 8; pg. 6 top left.

In the same work, we learn that, for Aristotle, the phrase, "present in a subject", means "incapable of existence apart from the said subject".

> By being 'present in a subject' I do not mean present as parts are present in a whole, but being incapable of existence apart from the said subject.
> ——*Categories*; Chap. 2; 1a: 22-23. **Great Books**; Vol. 8; pg. 5 bottom left.[41]

For Aristotle, then, all substrata are "substance", and the terms "substratum" and "substance as substratum" mean "that which is neither predicable of a subject (*i.e.:* stratum) nor incapable of existence apart from a subject".

Now comes an important question. In the Aristotelian system, is matter substance? Here is where all the confusion begins. Aristotle, you see, both specifically affirms it and just as specifically denies it. Thus, we read:

> If we adopt this point of view, then, it follows that matter is substance. But this is impossible; for both separability and 'thisness' are thought to belong chiefly to substance.

---

[41] **NOTE OF APRIL 8, 2002:** I repeat: For myself (and, in my opinion, for Aristotle as well), the phrase "present in a subject" means incapable of avoiding annihilation apart from *intimate bonding* to the said subject. What is "present in" a subject *coheres* to that subject and sticks to it so dependently that, the instant the bond to the said subject is broken, it necessarily lapses into nothingness leaving behind not the slightest remnant of itself.

——*Metaphysics*; Book VII; Chap. 3; 1029a: 27-28. *Great Books*; Vol. 8; pg. 551 bottom right.

> But clearly matter also is substance; for in all the opposite changes that occur there is something which underlies the changes.
> ——*Metaphysics*; Book VIII; Chap. 1; 1042a: 33-34. *Great Books*; Vol. 8; pg. 566 upper right.

To remove the confusion, one must first realize that, for Aristotle, "substance" means far more than one mind can usually tolerate. In the first place, "substance" means the same as "hypostasis". That is to say, it means a single object, such as a man, which is undivided within itself and divided from all others. As such, substance is a "concrete thing".

> . . . for the definition is a single formula and a formula of substance, so that it must be a formula of some one thing; for substance means a 'one' and a 'this', as we maintain.
> ——*Metaphysics*; Book VII; Chap. 12; 1037b: 25-27. *Great Books*; Vol. 8; pg. 561 top right.

Besides meaning "the concrete thing", for Aristotle, "substance" also means the "formula" which, taken with the matter of the concrete thing, produces the concrete thing.

Since substance is of two kinds, the concrete thing and the formula (I mean that one kind of substance is the formula taken with the matter, while another kind is the formula in its generality), substances in the former sense are capable of destruction.
——*Metaphysics*; Book VII; Chap. 15; 1039b: 20-23. *Great Books*; Vol. 8; pg. 563 mid right.

As "concrete thing", "substance" includes matter, form, and the composite of the two, and each of those three is also "substance".

But let us resume the discussion of the generally recognized substances. These are the sensible substances, and sensible substances all have matter. The substratum is substance, and this is in one sense the matter (and by matter I mean that which, not being a 'this' actually, is potentially a 'this'), and in another sense the formula or shape (that which being a 'this' can be separately formulated), and thirdly the complex of these two, which alone is generated and destroyed, and is, without qualification, capable of separate existence; for of substances completely expressible in a formula some are separable and some are not.
——*Metaphysics*; Book VIII; Chap. 1; 1042a: 24-32. *Great Books*; Vol. 8; pg. 566 bottom left and top right.

> And why is this individual thing, or this body having this form, a man? Therefore what we seek is the cause, i.e. the form, by reason of which the matter is some definite thing; and this is the substance of the thing.
> ——*Metaphysics*; Book VII; Chap. 17; 1041b: 6-8. *Great Books*; Vol. 8; pg. 565 mid right.

We must also bear in mind that, in addition to the formula, the concrete thing, the matter, the form, and the composite, there is yet the essence and the universal, and they, too, are "substance".

> Let us return to the subject of our inquiry, which is substance. As the substratum and the essence and the compound of these are called substance, so also is the universal.
> ——*Metaphysics*; Book VII; Chap. 13; 1038b: 1-3. *Great Books*; Vol. 8; pg. 562 left.

To extricate us from this impossible maze, Aristotle makes his famous distinction between the potential and the actual. Though the following two quotes may not answer all our questions about what "substance" means to Aristotle, they at least tell us that matter is *potentially* substance and substratum without *actually* being such, and that explains how the matter *is* the substance in one sense while it *is not* the substance in another sense. That will give us all we need for our purposes.

Clearly, then, if people proceed thus in their usual manner of definition and speech, they cannot explain and solve the difficulty. But if, as we say, one element is matter and another is form, and one is potentially and the other actually, the question will no longer be thought a difficulty.

——*Metaphysic*; Book VIII; Chap. 6; 1045a: 20-25. *Great Books*; Vol. 8; pg. 570 upper left.

But, as has been said, the proximate matter and the form are one and the same thing, the one potentially, and the other actually.

——*Metaphysics*; Book VIII; Chap. 6; 1045b: 18. *Great Books*; Vol. 8; pg. 570 mid right.

After all the verbal display is done, and the fine distinction between the actual and the potential is drawn, we still come back to the fact that, for Aristotle, as for virtually every reasonably intelligent man, matter is substratum and substance. Since, for the purposes of this dissertation, substance as substratum is our only concern, let us now ignore all the other meanings Aristotle attaches to "substance", and let us concern ourselves with what he says about substrata.

Very well, what *does* Aristotle say about substrata? I say that nowhere does Aristotle give us a thorough analysis of the idea of substrata. He says virtually nothing about it. For one thing, the only kind of substrata ever mentioned by Aristotle is substance, and matter is clearly included in substance. The only other thing we are told is that "substance", "substratum", and "matter" mean "that which is _neither_ predicable of a subject _nor_ incapable of existence apart from a subject". That, I say, is the sum total of the Aristotelian doctrine on substrata.

Having said that, I will hasten to assert, nevertheless, that there is probably one other idea which we can _read into_ Aristotle's idea of substrata. For all his discussion of the potential and the actual, I dare say that, for Aristotle, substance, as substratum, is homogeneous, uniform, and unalloyed. In other words, for Aristotle, every iota of a particular object's substratum is exactly like every _other_ iota of its substratum.[42]

At this point, let us define "substantial substratum" as: "any substratum which is neither predicable of another substratum nor incapable of existence apart from a subject". With that definition in hand, we can now state the Aristotelian position on substrata in the following three propositions:

(1)   All substrata is substantial substrata.

(2)   Matter is substantial substrata.

(3)   Every unit of substrata is, within itself, absolutely uniform.[43]

---

[42] **NOTE OF APRIL 8, 2002:** In other words, for Aristotle, every iota of something's substratum is of the same, one kind—namely: that which is neither present in nor predicable of another.

[43] **NOTE OF APRIL 8, 2002:** In other words, for Aristotle, every unit of substrata is, throughout itself, that which is neither present in nor predicable of another.

SUBSTRATA

Now comes a crucial question. What happens if Aristotle's position is in error? Well, I think it rather evident that, at least in Aristotle's case, his ideas regarding "being" are inextricably bound up with his ideas regarding substance as substrata. At the same time, it is generally conceded that Ontology, the study of being, is the starting point of philosophy. If all of that is true, then, the starting point and corner stone of the whole Aristotelian system rests on Aristotle's notion that all substrata is substantial. Naturally, if Aristotle's notion of substrata is in error, then the whole Aristotelian system is but a house built on sand, and the same will be true of any other system founded on the same, or basically similar, ideas regarding substrata.

By now, dear reader, you are probably saying: "Aha! You are proposing that Aristotle's conception of substance as substrata is a fallacy."

I am saying no such thing, dear reader. It is utterly unnecessary to prove the slightest error in Aristotle's conception of substance and substrata. It is unnecessary for the simple reason that Aristotle nowhere presents any arguments or evidences to support his notions regarding substance and substrata. Why should I bother to assault or to refute what no one ever bothered to prove?

Aristotle's notions regarding substance as substrata are nothing but totally gratuitous assumptions. Neither he nor anyone else has ever shown why we must assume either that all substrata is uniform and substantial or that matter is a uniform, substantial substrata. Nevertheless, Aristotle was quite right in making those gratuitous assumptions and was quite right in not bothering to prove them, because, till now, no one had ever come up with an even remotely viable definition of *in*substantial substrata and *non*-uniform substrata, and certain it is that no one needs to prove that for which there is manifestly no known alternative.

On the other hand, in my opinion, Aristotle, as well as all those who copied and copy his starting point, definitely took it for granted, and still take it for granted, that no alternative to the Aristotelian starting point has ever been successfully promoted precisely because no definition is possible of insubstantial and non-uniform substrata. *That*, I contend . . . *that* is their fundamental fallacy. There *is* such a thing as non-uniform substrata and insubstantial substrata, and their definition *is* possible. *That*, I will now prove by presenting such definitions, and that will be my alternative to the Aristotelian starting point.[44]

What I am about to set forth here is nothing more than a series of gratuitous assertions. Since, however, Aristotle's starting point is also nothing more than a series of gratuitous assertions, my starting point necessarily has as much validity as his does. In this department, I, like everyone else, can think whatever I care to think. As long as my ideas are reasonable and relevant, each and every devotee of philosophy will have as much right to choose mine and to reject Aristotle's starting point as he has to choose Aristotle's and to re-

---

[44] **NOTE OF APRIL 8, 2002:** The meaning of the above might be clearer, if we add this question: Is every iota of the substrata within a particular object (such as this horse or that man) of that kind describable as "that which is neither present in nor predicable of another"? If you answer yes, then you're advocating *uniform* substrata. If you answer no, then you're advocating *non*-uniform substrata. To put it another way, if you answer yes, you're asserting that all objects are uniform with regard to their substrata, because all the substrata in every object are of the same one kind—namely substantial.

ject mine, since neither side offers anything more to support its opening ideas other than their reasonableness and their relevance.

I will also add this: If a viable alternative to the Aristotelian starting point is in fact presented herein, then no man shall hereafter call himself a philosopher unless he is well aware of what is said here of those two starting points and their differences. Let us, therefore, proceed to the elucidation of them.

(B)

To begin with, let us define substrata. I define substrata as "that which is not predicable of a subject". That is not quite the same as Aristotle's definition, since he defines it as "that which is neither predicable of a subject nor incapable of existence apart from a subject". Before moving on, let me point out one thing in passing. The definition I have just given is merely one I give for the purposes of this article. My own, personally preferred definition of "substrata" is: "that which either is or can be involved in activity".

There are two genera of substrata. The first genus is "substantial substrata", and the second genus is "insubstantial substrata". The difference between the two genera is this: Substantial substrata is either actually or potentially existent apart from all other substrata, while _in_substantial substrata is neither actually nor potentially existent apart from substantial substrata.[45]

---

[45] **NOTE OF APRIL 8, 2002:** Note that the reference is to the _genus_ "substantial substrata" and the _genus_ "insubstantial substrata". In effect, I was saying: "The definition of 'substan-

By the phrase, "apart from", I do not mean "without any kind of assistance from". Like Aristotle, I mean only "is not present in".[46]

How does this difference between substantial and insubstantial substratum come to pass? To answer that, we must delve into the diunity of finite substance. Note that I said *di*unity and not *dis*unity.

First of all, I define substance as "that which is either actually or potentially existent apart from all other substrata".[47] At least, so I define it for the pur-

---

tial substrata' is: 'either actually or potentially existent apart from all other substrata', and the definition of 'insubstantial substrata' is: 'neither actually nor potentially existent apart from substantial substrata'." Clearly, it's a circumstance demanding the singular rather than the plural form of the verb "to be".

[46] **NOTE OF APRIL 8, 2002:** Be doubly assured of it: The above sentence is indeed found in the original version of *Aristotle's Fundamental Fallacy*.

[47] **NOTE OF APRIL 23, 2002:** As I wrote that definition, what I had in mind was ultimate constituents rather than the material things we observe such as this particular man or that particular horse. As a result, I was thinking in the terms of indivisible units of substantial substrata each of which is in no way "stuck" to any other such indivisible unit. At the same time, though, I was thinking in the terms of finite forms and generators—units in which an *actually* substantial substratum, of a definite quantity, is "stuck" to a *potentially* substantial substratum of a definite quantity. "Infinite substance", then, signified an indivisible, actually substantial substratum which, as such, actually exists apart from intimate bonding to all other substrata *whatsoever*, and "finite substance" signified an indivisible, actually substantial substratum which, as such, actually exists apart from intimate bonding to all other units of indivisible, actually substantial

poses of this paper. Elsewhere, I define it as "that which either is, or can be, involved in activity after the manner of an agent", and to be involved in activity after the manner of an agent is to be the doer of the activity in which one is involved, which is the same as saying one is *actively* involved in one's activity.

Next, I gratuitously deny that finite substance is homogenous. Finite substance is never uniform and unalloyed.[48] I gratuitously assert that, as substantial

---

substrata, but, simultaneously, does not by any means exist apart from intimate bonding to all other substrata *whatsoever*. After all, because it is finite, it is intimately bonded to its own potency. In defining "substance" as "that which is either actually or potentially existent apart from all other substrata", I definitely covered both infinite and finite substances. For, in *infinite* substance, the substantial substratum *actually* exists without cohering to any other substratum *whatsoever*, and, in *finite* substance, the substantial substratum has the *potential* to cease cohering to its potency and, upon fully using that potential (something only God can cause it to do) will exist without cohering to any other substratum *whatsoever*. Unfortunately, since most readers would probably think of "substance" in the terms of men and horses rather than indivisible ultimates, it's not very likely my definition would make any sense to many. The way I think, men, horses, etc., are not "substance" except from the standpoint of what I call their "leading form". For, in each and every man, horse, etc., the leading form is indeed an indivisible ultimate. At least, that's the way it is in *my* philosophical system. Even in Aristotle himself, there's some precedent for the idea of substance as indivisible. See my quote from him on pages 288 & 289.

[48] **NOTE OF MAY 1, 2002:** Typing that, I had forms and generators in mind. Since each *can* exist apart from all other substrata, each is "substance" as defined in the prior paragraph. If every form and generator is "substance", substance is by no means uniform and unalloyed. For, forms encase 2

substratum, finite substance is at least intrinsically di-une. In other words, if it were possible to come face to face with an indivisible unit of finite substantial sub-strata, one would come face to face with one thing; and yet, it would have two distinct "faces" (*i.e.:* essences), and each of these two essences (*i.e.:* "faces") would be radically different one from the other to an absolutely astounding extent, since one is *active* and the other *passive*; one is *dynamic* and the other *static*; one is *alive* and the other *frozen*.[49]

---

and generators 3 kinds of substrata. Unfortunately, it's a bad use of the term "substance". I should not have used "substance" at all.

[49] **NOTE OF APRIL 8, 2002:** I should probably have said this: "If it were possible to come face to face with an indivisible unit of finite *actually* substantial substrata, one would find that it is intimately and inextricably bonded to a unit of *potentially* substantial substrata. Necessarily, that would produce a single set of unbroken extremities and, therefore, what is *one* thing from the standpoint of that set of unbroken extremities remains *two* things from the standpoint of what is enclosed within that set of extremities. You would thus have one *object* encompassing two *subjects* (*i.e.:* two radically different kinds of substrata) each of which has a 'face' and an essence radically different from that of the other." Notice how, in the next paragraph above, I referred to the "one thing" as "one 'lump' of substratum". That should have been a dead give-away that I was talking of what is one from the standpoint of a set of extremities not intimately bonded to any other set, but two things from the standpoint of what that set of extremities encloses. Still, I should not then have said: "each unit of finite substance is two forms of reality which cannot be separated one from the other, except in thought." I should have said: "each unit of finite substance is a unit of actually substantial substrata so intimately bonded to a unit of potentially substantial substrata as to produce a kind of

SUBSTRATA

In other words, while being one thing and one "lump" of substratum, each unit of finite substance is two forms of reality which cannot be separated one from the other, except in thought. I gratuitously assert that each unit of finite substance (*i.e.:* each "lump" of finite substantial substratum) is an indestructible, indissoluble, ontological union of a negatively (*i.e.:* passively) substantial substratum with a positively (*i.e.:* actively) substantial substratum.

"Positively substantial substratum", "actively substantial substratum", "actually substantial substratum", and "*actuality*" mean the same thing. For the purposes of this paper, I define them as "that species of substantial substratum which is currently being *not present in* any other substratum".[50]

"Negatively substantial substratum", "passively substantial substratum", "potentially substantial substratum", and "*potency*" mean the same. For the purposes of this paper, I define them as "that species of

---

reality (*i.e.:* an object) encompassing two other kinds of reality which cannot be separated from one another, except in thought." In short, I must admit that, in the above paragraphs, I didn't do a very good job of expressing my thoughts. Still, I think that what I said was sufficiently clear to allow others to get my drift. Certainly, it was clear enough that men with an alphabet soup behind their names should not have misinterpreted it as grossly as did Mr. Butts, Fr. Clarke, and Mr. Scaltsas. At least, so say I. History may decide differently, and then God will make the absolutely final judgment. I may take issue with what *history* says; but, I have no quarrel with the decisions of 𝒯he 𝒜nfinitely 𝒜nformed.

[50] My preferred definition is: "that species of substantial substratum which is currently being the agent (*i.e.:* the doer) of an act of existence" or "what *is* actively involved in activity".

substantial substratum which, while *currently* it is only *present in* a positively substantial substratum, *can become* actuality".[51]

That is the same as saying that, in the system set forth here, "potency" is no mere abstract concept. It is a real "something" ontologically joined to the actuality of substance and serving as the reserve from which substance draws more and more actuality. Such an idea is by no means *totally* without precedent in Aristotle. In his **Metaphysics**, he writes:

> . . . for each thing is a unity, and the potential and the actual are somehow one.
> ——**Metaphysics**; Book VIII; Chap. 6; 1045b: 20. **Great Books**; Vol. 8; pg. 570 right just below the middle.

For the sake of clarity, let us attempt to express the diunity of substance another way. Let us say this: Every unit of substance is simultaneously and necessarily being both itself and the mirrored image of itself. So I gratuitously assert, and gratuitously add that, in the realm of primary principles, a thing and its mirrored image are not equally real. Thus, while being one substantial substratum, a unit of substance is being two mutually reverse forms of reality. In the "positivity", "activeness", and actuality of itself, the unit of substance is *actually* being what it is currently being. Simultaneously, in the "negativity", "passiveness", and po-

---

[51] My preferred definition is: "that species of substantial substratum which, while it is currently adhering in, participating in, and receiving a positively substantial substratum's act of existence, can be converted into actuality", or "what *is* passively involved in activity but *can be* actively involved".

tency of itself, it is *really* being all that it will ever be in the future, but it is not currently being it *actually*, *positively*, and *actively*; it is being it *potentially*, *negatively*, and *passively*.[52]

Nevertheless, in this system, every unit of substance does contain, within its current, positively real self, a negatively real, mirrored image of its future self, and this reversed, anti-image of its future self (*i.e.:* this negatively real aspect of an intrinsic diunity) must, of necessity, affect the positive side of that diunity. How shall it affect it?[53] I gratuitously assert that it shall act as the drag which "slows down", retards, delimits, and prevents the unit of substance from immediately and

---

[52] **NOTE OF APRIL 8, 2002:** The above paragraph should read so: In the act of being itself, every unit of substance is simultaneously and necessarily being intimately bonded to the mirrored image of itself. So I gratuitously assert, and gratuitously add that, in the realm of primary principles, a thing and its mirrored image are not equally real. Thus, while being one substantial substratum, a unit of substance produces an *object* encompassing two mutually reverse forms of reality. In the "positivity", "activeness", and actuality of the *object*, the unit of substance is *actually* being what it is currently being. Simultaneously, in the "negativity", "passiveness", and potency of the *object*, the substance's mirrored self is *really* being all that the substance itself will ever be in the future. The mirrored image is not being it *actually*, *positively*, and *actively*; but, it is being it *potentially*, *negatively*, and *passively*.

[53] **NOTE OF APRIL 8, 2002:** Change the above sentence to read so: Nevertheless, in this system, every unit of substance does hold—intimately and inextricably bonded to its current, positively real self—a negatively real, mirrored image of its future self, and this reversed, anti-image of its future self (*i.e.:* this negatively real aspect of an intrinsically diune *object*) must, of necessity, affect the positive side of that diunity.

currently being all that it could be if it could convert all of its potency to actuality in one act.

What is the effect of this drag? I gratuitously assert that the result is a line of tension between the actuality and the potency of the unit of substance. That is to say the unit of substance—by acting in the presence of a drag inducing differential between the actuality and the potency of itself—generates within its actuality, and by means of its potency, a feelable line of force having two termina—one of which is the actuality of the unit of substance, while the other is the potency of that unit.

Examine this "feelable line of force" which is extended between the actuality and the potency of the unit of substance. Is it a positively substantial substratum? One can hardly say that. It is merely a kind of "disturbance in the ether", if one may be permitted to use an outdated imagery. That "track of disturbance" is being generated in "the ether" (*i.e.:* in the positively substantial substratum) by "the ether" itself but *by means of* the "ether's" own anti-self (*viz.:* the negatively substantial substratum). Is it, then, a negatively substantial substratum? How could we say that? It is by no means something describable as a mirrored, reversed form of anti-reality waiting and able to be converted into actuality. What, then, is it? I gratuitously assert it can only be a third kind of substratum which we shall call "insubstantial substratum", and, for the purposes of this paper, insubstantial substratum is defined as "that genus of substrata which is never, whether actually or potentially, not present in another substratum".[54]

---

[54] My preferred definition is: "that genus of substrata which is neither actually nor potentially the agent of an act of existence, but which is always limited to being the patient of one act of an agent.

SUBSTRATA

[55]For the starting point set forth here, then, finite substance—as the smallest possible unit of substratum to which anything is potentially divisible—is intrinsically triune. While being one thing, each unit of finite substance necessarily includes a positively substantial substratum, a negatively substantial substratum, and a third substratum which is neither one of the other two species of substantial substrata, since it is an insubstantial substratum extended between the other two. That means that, if we could "come face to face" with what each of those three substrata is being in itself as a substratum (*i.e.:* if we could "see" the essence of each)[56], we would, while "seeing" one thing, observe three essences each of which is most radically different from the other two.[57]

---

[55] **NOTE OF APRIL 8, 2002:** Change the first 2 sentences of the above paragraph to read: For the starting point set forth here, then, finite substance—as the smallest unit of substratum to which anything is divisible—always and necessarily produces a triune object. While being one thing as a set of unbroken extremities, each object necessarily includes a positively *substantial* substratum, a negatively *substantial* substratum, and a third substratum which is neither one of the other two species of *substantial* substrata, since it is an *in*substantial substratum extended between the other two.

[56] It is my contention that, outside of the beatific vision, we can only come face to face with the essence of the lines of force (*i.e.:* the insubstantial substratum), and that we do whenever we see a sight, hear a sound, smell an odor, taste a flavor, or feel a tactile sensation.

[57] Therefore, from the standpoint of *what* it is in itself (*i.e.:* from the standpoint of its essence), each substratum is wholly "other than" the other two, and, *in that sense*, each is entirely "not present in" the other two. Nevertheless, from the stand-

To summarize this section, then, there are two genera of substrata. The first genus is substantial substratum, and it includes the species actually substantial substrata and potentially substantial substrata. The second genus is insubstantial substrata, and it contains only one member. Insubstantial substrata, therefore, is "*sui generis*".

Actually substantial substrata can be called a "being in itself". Potentially substantial substrata can be called a "being included in another". Insubstantial substrata can be called a "being generated and extended between two others that are one".

Finally, Aristotle gave us only one kind of substrata (**viz.:** that which is neither predicable of another substratum nor present in another substratum). We now have one kind of substratum *actually* not present in another, one kind of substratum *potentially* not present in another, and a third kind of substratum necessarily present in between the two species of a second genus of substrata. A Most remarkable difference, wouldn't you say?

(C)

Aristotle told us that the question, concerning what being is, is the same as the question concerning what substance is. If that is true; and if substance, as the ultimate substrata (Let us be courteous to Aristotle and say *potentially* ultimate substrata.), is intrinsically triune—then it follows that being is also intrinsically

---

point of *how* it is what it is (*i.e.:* from the standpoint of its *being*), the second substratum is *in* the first, while the third is *in between* the second and third substrata.

three fold. Therefore, let us define the two species of substantial being and the one member of the genus insubstantial being.

Being is involvement in activity. The two genera of *being* are substantial being and insubstantial being. Substantial being includes the two species actual being and potential being.

"Actual being", "active being", and "positive being" are equivalent terms. Active being is the first principle of activity of that which is involved in activity after the manner of an agent.

"Potential being", "passive being", and "negative being" are equivalent terms. Passive being is the first principle of activity of that which is involved in activity after the manner of a patient waiting and able to be an agent.

The genus insubstantial being has only one member, viz. neutral being. Neutral being is the first principle of activity of that which is involved in activity after the manner of a patient which can never be an agent. I sometimes call neutral being "accidental being".

(D)

From what we have just said above, "being" is merely the most basic and first method of involvement in activity, which is to say "being" is "activity" and, therefore, "act". But, if being is three fold; and if being is activity—then it follows that there should be three kinds of activity, which is to say three kinds of acts. Furthermore, from what we've said about substance and being, it would seem there should here, too, be two kinds of substantial activity and one kind of insubstantial activity which is neither one of the two kinds of substantial activity.

To begin with, then, let us define substantial activity as that kind of activity which substance either _is_ doing or _can_ do. What are the two species of substantial activity which substance either is doing or can do?

We have been speaking of a unit of substance in which there is potency waiting to become actuality.[58]

---

[58] **NOTE OF APRIL 30, 2002:** Change the above sentence to read: We have been speaking of a unit of substance which is intimately and inextricably bonded to a potency waiting to become actuality.

Possibly, much of this confusion would have been avoided had I defined "substance" as "object". Substances would then be either reducible or irreducible objects (**_i.e.:_** compound or simple objects), which is to say objects which can be broken apart into 2 or more objects vs. those which cannot. Irreducible substances would then be either "mono-ousious" or "poly-ousious", which is to say objects whose extremities encase only one kind of subject (**_i.e.:_** substratum) vs. those whose extremities encase more than one. Poly-ousious objects would then be either di-ousious or tri-ousious, which is to say objects whose extremities encase _two_ kinds of inseparable subjects (**_viz:_** potency and actuality) vs. those which encase _three_ (**_viz.:_** potency, actuality, and a line of force). Every reducible object ultimately reduces to 2 or more tri-ousious, irreducible objects. There is, though, one problem in equating "substance" and "object": It leads to the conclusion that whatever is encased within an object is a _part_ of a _substance_, and that's a bum conclusion. In every _reducible_ object, each of the _ir_reducible objects is there as a _co-tenant_ of the _object_ and an _adjunct_ to the _substance_. In every tri-ousious irreducible object, the potency and the line of force are there as _co-tenants_ of the _object_ and _adjuncts_ to the _substance_. In every di-ousious irreducible object, the potency is there as a _co-tenant_ of the _object_ and an _adjunct_ to the _substance_. Maybe "substance" should mean "the dominant co-

SUBSTRATA

Can there be a substance in which all the potency is already developed into actuality? If there is such a unit of substance, its activity would be "pure activity", "pure actuality", and "pure act". Such a unit of substance would best be called "infinite substance", and its activity is "pure actuality". Pure actuality, therefore, is the first species of substantial activity, and it is the activity of that kind of substantial substratum which has no real _un_developed potency.

We have already seen the second species of substantial activity. It is impure activity. It is the activity of finite substance. It is the activity of that kind of substantial substratum which has not yet converted all of its real potency (*i.e.:* negatively real) into actuality.

What is the third kind of activity? That is to say, what is the one member of the genus insubstantial activity? To answer that question, let us first clarify the nature of substantial activity.

Consider first the pure activity of infinite substance. Is it a series of acts, or is it one act? If it is a series of acts, what is the distinction between act and act? If, as we have said, the activity is pure, then it is unchanging. Throughout the activity, it is all actuality and no real potency. In other words, the ratio of actuality to potency is perpetually 1/0. Therefore, it must be that, for infinite substance, substantial activity is a single act the essence of which is one, enduring, fixed ratio of actuality to potency.

---

tenant of an object" and "object" mean "a substance and its crucial adjuncts". For Aristotle, whether "substance" or "substance and its adjuncts", all is uniform and unalloyed. For, in *either* case, there's only one kind of substratum. For myself, though, "substance and its adjuncts" includes 3 in the case of men, horses, etc..

Note well what we have just implied about an *act* of substantial activity. We have implied that, for infinite substance, an act of substantial activity is an *enduring* one which, from its inception (if it had one) to its end (if it will have one) is *defined* and *formulated by* a *fixed ratio* of actuality to potency. If such is what an act of substantial activity is to infinite substance, why should it not also be such for finite substance?

If the above is true, we can now say this: For a unit of finite substance, substantial activity is a series of impure acts each of which has a duration and each of which is *defined by, formulated by*, and lasts as long as the act is characterized by, the same, one, particular fixed ratio of actuality to potency. In short, the essence of an act of impure substantial activity is the duration of a particular, fixed ratio of actuality to potency.

Clearly, that implies this: A finite substance has acted as many times as its ratio of actuality to potency has changed. From that, it is a small jump indeed to say that, for finite substance, the ratio of actuality to potency varies directly as the number of times the substance has acted.

Now let us see if we can return to the original question. What is insubstantial activity? Well, imagine a unit of finite substance performing its first substantial act. The ratio of actuality to potency is 1/1. Fifty percent of its substantial substratum is actuality, and the other fifty percent is potency. How long does that last? Unless we can compare its duration to that of another unit, who can say?

Imagine now that our unit is performing its second substantial act. The ratio of actuality to potency is now 2/1. Two-thirds of its substantial substratum is actuality, and the other one-third is potency. There has been a change from one act to the only second act which can logically follow it. At least, so it is, if, as we said, the ratio of actuality to potency varies directly as the number of times the substance has acted.

Consider that change from one substantial act to the next. There was an instant when the one ratio ceased, and there was an instant when the second ratio commenced. Consider the moment between those two instants. Did it have any duration? If it did, then the substantial activity of the unit of substance is broken. Indeed, if the break had any duration, then there was an enduring moment when the unit of substance was not. Therefore, it must be that the moment of break and change (*i.e.:* the moment between the one moment when the one ratio quit and the second moment when the second ratio commenced) was without duration.

If dear reader, you have followed the above, you perhaps now perceive a chain of three successive moments: (1) an enduring moment characterized by a particular fixed ratio of actuality to potency; (2) a durationless moment of change; and (3) an enduring moment characterized by a different, particular fixed ratio of actuality to potency.

Consider that "durationless moment of change". Is not change an activity? Is it a substantial activity? Is it something which the substance did? You will probably say it was but I say you are wrong. Admittedly, the unit of substance did the act immediately *before* the change, and the unit of substance did the act immediately *after* the change, and the *result* was a change, but the *change itself* is not a third distinct act performed by the substance.

Here, then, is our third kind of activity. This is the activity which we call the sole member of that second genus of activity called "insubstantial activity", and we define insubstantial activity as the change of a particular unit of finite substance from one impure act to one of its logical consequents.

If we wish to be more exact in our definition, we can say this: Insubstantial activity is any succession of changes by a particular unit of finite substance from impure act to logically consequent impure act. In that case, we must point out that things are "in succession" when there is nothing of their own kind intermediate between them, and, between change and change, there is nothing "of their own kind" intermediate between them, since, between each durationless change, there is only enduring changelessness.

(E)

At the end of the second paragraph above, we said that insubstantial activity is the change of a particular unit of finite substance from one impure act to *one of* its logical consequent_s_. *One* of its logical consequent_s_???!!! From what we said, should there not be only *one* logical consequent to each impure act?

Were it not for the complicated nature of potency, that would be true. Let us examine that complicated nature of potency and see what it implies.

For Aristotle, every "concrete substance" (*i.e.:* "concrete thing") is, as substratum, *potentially* divisible "*ad infinitum*"—a frightening prospect, indeed, for modern physicists. In direct contradiction, we are saying that substrata can be divided only to approximately 2^{-256} times the diameter of the hydrogen atom before one reaches an indivisible unit of substrata. Aristotle, I suggest, would insist that the substratum is homogeneous. In direct contradiction to that, we are saying that every ultimate, indivisible unit of substrata is—as *sub-*

stantial substratum—intrinsically diune, having its potency as a form of negative reality only logically separable from the positive reality of its actuality. Now, we complicate the matter several times over by saying that the potency is, in fact, six potencies in one. How can that be? Let us examine that issue.

We have proposed that finite substance—*as a unit*—is intrinsically triune. It is triune because there is an opposition of undeveloped potency to actuality resulting in a line of tension between the potency and the actuality. By contrast, we said that infinite substance contains no undeveloped potency. That would seem to indicate that, in infinite substance, there is but one, absolutely uniform, unalloyed substratum.

That seems logically appealing; however, it is no less logically appealing to assume that what is universally intrinsically triune in the finite world will somehow maintain that intrinsic triunity even in infinity. Exactly how that triunity manages to persist in infinity is apparently impenetrable by logic and is certainly irrelevant to this dissertation; nevertheless, as we shall see, it is certainly very relevant to say that, even in infinity, substance maintains its three reference points in its one self.

Every impure act of every finite substance is clearly an act in potency to, and working toward, a pure act. But, in being in potency to pure act, finite substance is in potency to an infinite triunity. Look at that closely enough, and we will see this: In being in potency to an infinity which embraces three reference points in its one self, finite substance is in potency to infinity six different ways. To understand how that comes about, consider illustration #1 on the next page.

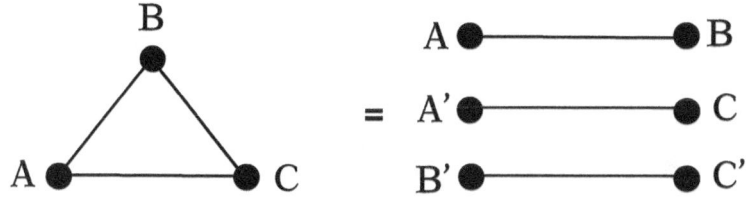

ILLUSTRATION #1

In illustration #1, we have three reference points: A, B, and C. Since each of the three reference points can be related to two other reference points, that gives us a total of six relationships. More specifically, it gives us three genera of relationships each embracing two mutually opposing species of relationships. In other words, AB and BA are a set of mutually opposing relationships. The same is true of AC/CA and BC/CB. Thus, we have three frames of reference each composed of two opposing poles. At least, so we have expressed it, for simplicity's sake, in our graphic illustration #1.

For simplicity's sake, then, the acts of finite substance can be conceived of as being related to three sets each having two opposite poles. I have attempted to re-express that even more simply in illustration #2 on the next page; but, since I cannot draw a three dimensional picture on a sheet of paper, I can only ask my reader to observe illustration #2 with a great deal of patience. Both illustrations should be kept in view as we proceed through the next few paragraphs.

If the sheet is lying flat on a table, imagine B' directly above Z. Imagine C' directly below Z. They should be so placed above and below Z that: (1) a line drawn from C' to B' will pass through Z and be perpendicular to lines drawn from A to B and A' to C, and (2) Z is equidistant from all six points.

271

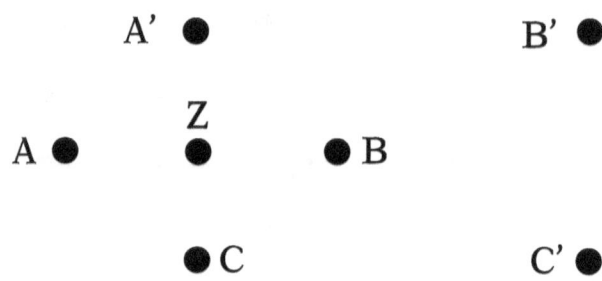

ILLUSTRATION #2

The point, A, represents A related to B. A' is A related to C. B is B related to A, and B' is B related to C. C is C related to A, and C' is C related to B.

The point, Z, represents a finite substance about to act for the first time. Since it has not yet converted any of its potency to actuality, it is equally in potency to all six reference points. For that reason, it is equidistant from all six. Naturally, Z is a seventh reference point.

Since there can be no such thing as a unit of substance which is all potency and no actuality (since potency cannot be unless it is being included in actuality), Z is merely an abstraction. Nevertheless, as an abstraction, Z can be said to have six ratios of actuality to potency, each of which is 0/1. Its "A ratio" is 0/1, because it has not yet performed a single act describable as "toward A and away from B". By the same token, its "B ratio" is 0/1, because it has not yet performed a single act describable as "toward B and away from A". We can say basically the same thing of its "A' ratio" and its "C ratio", and of its "C' ratio" and its "B' ratio".

There is, in fact, no reason to assume that a unit of substance develops its potencies one at a time. In all probability, the six potencies are developed three at a time with either a fixed or a floating ratio between the three ratios. Since, however, such questions are not within the purview of this work, let us imagine, for the moment, that each genus of ratios is developed one at a time.

Imagine Z acts for the first time, developing its "potency to B" into "act to B". As a result the "B ratio" becomes 1/1, and the opposite ratio, the "A ratio", becomes -1/1. From the standpoint of illustration #2, this first act means that Z will shift to the right along the line AB. How far will it move? If you're asking in the terms of inches, feet, centimeters, etc., I can't tell you. I can only tell you this: It has moved one act toward B, and, between Z and one act to B, there is no other act and, consequently, no other place. That means the only location immediately next to Z, on the way to B, is definable as: one act to B plus zero acts to all other points.

Imagine that Z acts ten acts developing its potency to B into act to B. Picture, if you wish, ten little dots in a straight line from Z to B. The B ratio is now 10/1, and the A ratio, its opposite, is -10/1.

Now, imagine this: As Z acts for the eleventh time, it develops its potency to A' instead of its potency to B. In that case, picture a little dot immediately above the last dot in the previous line of ten dots. The B ratio remains at 10/1 and the A ratio at -10/1. The A' ratio becomes 1/1, and its opposite, the C ratio, becomes -1/1.

Note that Z's line of motion was not directly toward the point A'. Its line of motion was, rather, parallel to a line drawn from A' to C and perpendicular to a line drawn from A to B.

Suppose, now, we decide to move to the left. We wish to go back in the direction of the original center and onward toward A. Our substance merely switches to developing its potency to A into act to A. With each successive act referenced to A, the A' and the C ratios remain at 1/1 and -1/1 respectively, while the B ratio diminishes, and the A ratio increases. Necessarily, that means Z's "act to B" will be reconverted into "potency to B" as the "potency to A" is converted into "act to A".

Again, note the line of motion. It is not directly to point A. It is, rather, along a line which is parallel to a line drawn from B to A. Z cannot pass through the original center point, since that would reduce all ratios to 0/1 and eliminate Z. The original center, then, is a "hole in space" which Z *must* avoid or perish.

If you have followed the above, dear reader, you can perhaps now perceive this: If the ratios are developed strictly one at a time, then there are always six logical consequents to each act, unless one is only one act from the original center. The six logical consequents are: one forward, one to the rear, one to the left, one to the right, one up, and one down. Of course, that is merely another way of saying that, unless we are adjacent to the center, the current substantial act can always be followed by a substantial act referenced either to A, to A', to B, to B', to C, or to C'. If, on the other hand, the ratios can be developed either one, two, or three at a time, then there are twenty-six logical consequents to each act. Furthermore, perhaps it is possible to develop two mutually opposite ratios simultaneously. That would result in two successive acts "in the same place", and that would be a twenty-seventh logical consequent.

But, let us not distract ourselves with too many new ideas. Let us concentrate, instead, upon the one notion which must be brought out here far more emphatically than any other. The long term result of what we are doing is this: We can easily construct a three-dimensional "dot-matrix" in which every dot's location can be defined, formulated, and fixed in the terms of a definite number of substantial acts referenced to each of three *generically* different reference points in infinity.

"So what?" you will ask. Well, remember what we have said. When a finite substance acts, it generates a line of force, and, because its potency is oriented six ways, that line of force will be oriented one of six ways. That means that, when substance acts, its act is applied to a well defined location in a three-dimensional matrix and "signs" that location with a line of force pointed in one of six directions. In other words, it starts drawing lines to connect the dots.

Conjure up any three-dimensional object you wish. We can easily reproduce it within our three-dimensional dot-matrix. All we have to do is to draw lines connecting the appropriate series of dots. Since we know the definition and formula for every dot in our three-dimensional matrix, we know exactly how many acts our units of substance must perform, in reference to which of the six infinite reference points, in order to accomplish the drawing.

Suppose now we want our drawing to "move through space". In that case, we use the same technique used by the motion picture studios to produce a "live action" cartoon. In other words, we draw the figure over and over again many times and, with each new drawing, introduce a slightly different location and, if the figure is an "animated" one, a slightly different attitude of the body and limbs. In our three-dimensional dot-matrix, that means that, with each new drawing of the physical object, the old drawing (*i.e.:* the lines of force generated by the *previous* acts of the substances) ceases to exist, and the substances now generate new lines of force connecting a slightly different complex of dots. Finally, if consciousness itself is an event which takes place in that durationless moment of change as some

275

substance capable of consciousness changes from actual act to actual act;[59] and if that substance experiencing the events of consciousness changes from actual act to actual act at a rate which is different from that of the substances generating the lines of force—then the result will be the same as that which we can observe every day on the screen of any motion picture theatre: A definite number of "still frames" is suddenly converted into a "motion picture", and the images on the screen seem to be enduring agents doing a series of "live actions". Nevertheless, in reality, the observed figures do not endure for more than one frame. They are merely projections passively sustained by another for one act, while it is the substances generating those figures which are *actually* moving.

That, then, is the explanation of how an _in_substantial substratum (*i.e.:* a substratum which manages to have being only because it is being generated within the diunity of a substantial host) can be the three-dimensional matter out of which everything in the physical universe is made. And when I say *everything*, I mean *everything*—including the brains in our heads and the sounds, flavors, odors, tactile sensations, and little colored shapes which run around in those brains, since, if matter is feelable "stuff", it can also be our sensations.

(F)

Here, then, is the *absolutely paramount* difference between the Aristotelian starting point and the al-

---

[59] **NOTE OF MAY 23, 2002:** I now consider that notion nonsense. Being a series of acts, consciousness must be *synonymous* with, rather than in *between*, the acts of a form.

ternative set forth here. For Aristotle, as for virtually every man with any meaningful degree of education and intelligence, the material objects of our physical universe are (*i.e.: potentially* are) what a *substantial* substratum is being, which is to say they are a substratum which is not present in another stratum. For Aristotle, consequently (and this is the more important factor), when we observe a physical object move from one place in space to another, we observe the substantial activity of a substantial substratum. Naturally, if motion through space is substantial activity, then motion through space must be absolutely continuous and absolutely unbroken, and time, since it is the measure of such motion, must also be absolutely continuous and absolutely unbroken.

The alternative set forth here, however, has clearly shown that it is most certainly possible to explain physical objects as something merely *generated* by substantial substrata and as something which an *in*substantial substratum is being, which is to say matter is a substratum present in between two strata which are one. More importantly, however, this alternative has shown what it means to say that motion through time and space is merely the result of the *in*substantial activity of substance. We have seen that, like insubstantial substrata, it is of the very essence of insubstantial activity to have being only as a result of being included "in between" that which is of another genus the very opposite of its own genus. In effect, we have seen that insubstantial substratum, insubstantial activity, motion through space, and time itself—all are of that genus of things whose very essence it is to fall in between the two species of an opposite genus. It is of their very essence, therefore, to be unbroken only in the sense that what falls between them must not be of the same species or genus.

That, after all, is why, between one durationless moment of time-and-change and the next durationless moment of time-and-change, there is only a timeless moment of duration-and-changelessness—a moment which, though it interrupts our world in one sense, nevertheless, does not interrupt it in the all important sense. That is so because, during that timeless moment of duration-and-changelessness, when a unit of substantial substratum is applying its substantial activity to the same, one place, there is no time; there is no change; there is no insubstantial activity; and there is nothing whatsoever of the things which belong to our physical universe. On the other hand, however, as often as there is a moment of *time* and _in_substantial *activity*, things are changing, in flux, and transitioning from place to place. So it is *as often as there is a moment of time and insubstantial activity*. For our world of insubstantial things, however, that is all that's relevant.[60]

Here, then, is the absolutely paramount difference between the Aristotelian starting point and the alternative one: Aristotle says the stainless steel ball is (*i.e.: potentially* is) a substantial substratum, and, if you throw it, the ball's movement through space will be the substantial activity of a substantial substratum. The al-

---

[60] One thing, nevertheless, should be mentioned in passing: If the duration of the timeless moments of substantial activity be increased, then the moments of durationless time will occur less frequently. Obviously, that means time is relative to the rate at which substance changes from actual act to actual act. The relativity of time is a well-established fact of observation these days. It is, however, utterly irreconcilable with Aristotle. As one can readily see, though, the relativity of time is perfectly in keeping with the alternative set forth here.

ternative, however, says the steel ball is an insubstantial substratum, and, if you throw it, the ball's movement through space will be the result of the insubstantial activity of the myriads of substantial substrata generating the insubstantial substratum being the ball.

Which position is the correct one? As I said earlier, the question is irrelevant to this dissertation, because Aristotle's position is an utterly gratuitous assertion which requires no refutation.

In saying that, I am not faulting Aristotle. Aristotle made the only assumption he could have made, because, prior to the twentieth century, the idea of an *in*substantial substratum (*i.e.:* a substratum *present in* between the potency and the actuality of the host substratum) did not and could not have existed, since only in modern times has the collective genius of the world's thinkers finally produced an environment in which the concept of an *in*substantial substratum has some *small* chance to be understood and to survive.

I say "some small chance", because, even now, it may not make it. It is now more than twenty years since I first attempted to communicate this idea to the world around me, and, to date, all I have accomplished is that my attempts to communicate this idea have been rejected by every philosophical journal on the North American continent, and I now face the very real prospect of dying before the idea can be understood by a single, other human being. If such is the predicament even today, how could Aristotle be expected to have foreseen that there could ever be another, radically different interpretation of the idea of "substratum".

Lest there be any grave misunderstanding here, let me hasten to point out one thing. The idea, that "force" or "power" is the "stuff" of which material objects are made, is by no means a new idea. It is commonly called "dynamism". That, however is not the new idea which this paper seeks to present. What is new here is the explanation of how a line of force manages to be a kind of substratum which is *present in* another substratum. The idea that a line of force is the "track of disturbance" generated in the actuality of a unit of substance when it acts in the presence of a drag inducing differential between the actuality and the potency of itself—*that idea* is what is new, and that idea is the concept which, till now, neither Aristotle nor any other philosopher, save this writer, ever grasped or even considered a possibility.

Leibnitz, Einstein, and others spoke of force as the "stuff" of which material objects are made. When, however, it came time to describe what they mean by force, their descriptions were invariably descriptions of a *substantial* substratum—of a stratum *not present in* another stratum. Their descriptions were such because they simply did not know how to think of a stratum present in another, and they knew not how to do that because no one had yet shown how a substratum could truly be a stratum and yet, simultaneously and as stratum, be incapable of existence apart from another substratum.

All of that, however, is now changed forever. For the first time in the history of mankind, an individual has explained to the world how it is possible to have a true substratum which must be present in another, and I am arrogant enough to boast that I have explained it most clearly, graphically, and thoroughly. As a result, it is now possible, for the first time in recorded history, for human beings truly to think the idea of a substratum which is present in another—which is *incapable* of existence apart from another.

Now, then, if all of that is true, certain it is that Aristotle cannot be faulted. It is equally certain in that

case, however, that his whole system is necessarily naked, unarmed, and helpless before an assault for which
he could not possibly have prepared it—a preparation
he could not make because he could not possibly have
foreseen the nature of the assault.

(G)

Earlier, we adverted in passing to another difference between the Aristotelian starting point and the alternative set forth here. It might be in order to expound
upon that difference a bit more fully before closing this
treatise.[61]

---

[61] **NOTE OF APRIL 9, 2002:** These days, much of what I said
in this section makes little sense even to me. What's said here
now strikes me as perhaps a very unsatisfactory attempt to
spell out and solve a problem which, probably, I have not
even now managed to grasp as fully as it needs to be grasped.
For many years, I have realized that there is an important
distinction between: (1) the characteristics internal to the actuality of a given form (*i.e.:* what one would observe could
one come face to face with a given form's actuality—something one can do only in infinity), and (2) the influence that
form has upon other forms and generators (*i.e.:* the capacity
of a form to *form* something composed of more than one ultimate constituent). Why harp on the distinction? It's because
it requires us to ask this: Is #2 the necessary result of #1? In
other words, could you observe the internal characteristics of
some form's actuality, would you, in doing so, observe all you
need to observe in order to predict precisely what kind of object that particular form would produce? To put it another
way: In observing the essence of some form's actuality, would
you also know the essence that form would produce in that of
which it is the form? I have long known that the answer is no,
because, to know precisely what object a form would form,

281

you must know a great deal about the pattern of divine rotation being followed by the heavenly sextets and/or benchmarks with which that form is associated. Necessarily, that means you must have much knowledge of something which is very much *outside* of the form. There is, therefore, an immense difference between: (1) the *act* whereby a given form's actuality acquires its own internal characteristics, and (2) the *activity* whereby a given form acquires its ability to fashion this or that particular material object, such as this man or that horse. The first (*i.e.:* the *act*) is the form's act of *existence*, which is to say its act of being an agent able to avoid annihilation without cohering to another. The second (*i.e.:* the *activity*) is the pattern of rotation executed by the form's actuality in response to the heavenly sextets and benchmarks with which it is associated. Its act of *existence*, then, is the *act* which determines what essence the form *itself* will have. On the other hand, the rotational activity imposed upon it by the heavenly sextets and benchmarks is the *activity* which determines what essence the form will impose on that of which it is the form; and so, the essence *of* the form (*i.e.:* the form's *own* essence) and the essence *imposed by* the form are as different from one another as the infinite is from the finite. This awesome dichotomy of "essence of the form itself" (*i.e.:* essence of the form's act of existence) vs. "essence imposed by the form" (a dichotomy which disappears only in God's Divine Actuality) is one which must be grasped, and the only way one can even begin to do so is to understand something about the different ways to be involved in activity. To be merely an *act* (*i.e.:* to be just another agent, non-inhering subject, and substantial substratum) with its own individuating essence, every form must be involved in *existence* (*i.e.:* it must be involved in an act of "being" and must be involved after the manner of an agent able to avoid annihilation even without being bonded to any other agent), and, to do that, every form must be in *potency* to infinity's act of existence; but, to be a *form* (*i.e.:* to be that particular kind of agent able to fashion a complex structure in which many agents are included), every form must be *associated* with a particular pat-

On page 242 of this essay, Aristotle told us that the question, regarding *what* "being" is, is the same as the question about *what* substance is. Put bluntly, he equated "being" with the essence (*i.e.:* the "whatness") of substance. Since he nowhere shows why we should make such an equation, this too is but a gratuitous assertion constituting part of his starting point.

The alternative says that "being" is involvement in activity. Consequently, rather than pertaining to *what* substrata is, "being" pertains to *how* substrata is what it is—pertains to *how* it is connected to activity. Thus, "being" is a principle of activity quite distinct from the principle of essence.

If essence and being are in fact two different principles, then we can rightfully give two different definitions of substrata—one from the standpoint of its essence, and the other from the standpoint of its "being".[62] Thus, from the standpoint of its essence, all substrata is *neither* predicable of a subject *nor* present in a

---

tern of divine rotation. As some would put it: Every form is the kind of act which remains in potency to an act of existence; but, it does so as merely an *agent* rather than as a *form*. As I myself prefer to express it: Every form is the kind of act which remains in *potency* to a *rectilinear act*; but, it does so as merely an *agent* rather than as a *form*. To be a *form*, it must be *associated* with *curved activity*. Forearmed with such thoughts, the reader can perhaps now return to the above and read without getting *hopelessly* lost.

[62] **NOTE OF APRIL 2, 2001:** To define a substratum "from the standpoint of its essence" is to define it from the standpoint of the results of its internal characteristics. To define a substratum "from the standpoint of its being" is to define it from the standpoint of the results of the way it is involved in activity.

subject,[63] since, *as an essence*, each substratum is strictly *an individual*. Notwithstanding that, from the standpoint of its "being", each substratum, while never *predicable* of a subject, *may* or *may not* be *present in* a subject. It all depends upon the species and genus of the stratum, because, as *"individuals* possessing being", some substrata are more *independent* individuals than are others.

For example, infinite substantial substrata is an *absolute agent*, because it is the doer of an act which it originated (*i.e.:* it is an *"agens in se a se"*); positively substantial substrata is a *relative agent*, because it is the doer of an act which it must always receive from another (*i.e.:* it is an *"agens in se ab altero"*); negatively substantial substrata is a *relative patient*, because *currently* it is merely "being *passively carried* on the back of" a relative agent's received act of being, as it awaits conversion into *active* participation in that act; and, finally insubstantial substrata is an *absolute patient*, because it can never participate *actively* in an act of being and is always limited to being a patient *passively* carried along for one act of a substance.

---

[63] **NOTE OF APRIL 2, 2002:** In the above case, of course, I'm using the phrase "present in" in the sense in which the parts and characteristics of a subject are "present in" their subject. Naturally, no subject—as long as it is still truly a subject—is ever "present in" another subject as either a part or characteristic of that other subject; otherwise, it would have no essence of its own and would merely be some aspect of the essence of something else. Nevertheless, some subjects—even while still truly being subjects—are "present in" another subject in the sense that they are incapable of existing apart from intimate bonding to that other subject.

For Aristotle, that is all gibberish, because, for him, whatever is being an *individual* is necessarily being a substantial substratum united with a form. Of course, if each and every *individual* thing is being in the same identical sense (**viz.:** being a substantial substratum), then it follows with absolute necessity that each and every individual is equally independent, and any system, which proposes to examine into the different degrees of independence to be found among *individuals*, is sheer sophistry. It is one thing, however, to *say* each and every individual thing is equally independent; it is quite another thing to *prove* it.

But, let us go on with the alternative. If we assume that being and essence are two different principles; and if we follow Aristotle's own logic—then it follows that an hypostasis (**i.e.:** an *individual* thing such as "this man" or "this horse") is a composite of *four* rather than *two* elements. Instead of matter and form, there is: (1) the matter, (2) the "being" of the matter, (3) the form, and (4) the "being" of the form.[64]

---

[64] **NOTE OF APRIL 9, 2002:** It is a composite of 4, if, like Aristotle, we think of visible material objects (such as this man or that horse) in the terms of matter and form only. In the "Haasian" system, one thinks of material objects in the terms of matter, form, and the generators of matter; and so, to the 4 principles given above, we add these 2: (5) the generators, and (6) the "being" of the generators. Indeed, to be *perfectly* exact, add even these other 3: (7) the potency of the form, (8) the potency of the generators, and (9) the "being" of those two potencies. Since the "being" of potency is the same in all cases, there's no need to list the potency of the form and the potency of the generators separately. Incidentally, though in St. Thomas' system there is only one form per visible material object, in the "Haasian" system, every such object encloses an astronomical number of forms. Still, there is only *one* _leading_ form. Well, in the "Haasian" system, the body it-

Immediately, that compels us to ask a very crucial question: Can there be such a thing as matter having a particular quantity, quality, place, position, state, affection, and relation as a result of its conjunction with an act of existence alone and apart from any conjunction with a form? But, what does that mean? Does it mean "being" is another kind of form? Do we, then, have a form and a form of the form? What, then, is the difference between form and "being"? Aren't we merely multiplying principles needlessly?[65]

---

self has a leading form; but, since it is subservient to the soul, the soul is *absolutely* the leading form.

[65]   Precisely because existence reaches substance in and through its form, forms have to receive existence in order that they become "being". But Thomas Aquinas could not posit existence (*esse*) as the act of a substance itself actualized by its form, without making a decision which, with respect to the metaphysics of Aristotle, was nothing less than a revolution. He had precisely to achieve the dissociation of the two notions of form and act. This is precisely what he has done and what probably remains, even today, the greatest contribution ever made by any single man to the science of being. Supreme in their own order, substantial forms remain the prime acts of their substances, but, though there be no form of the form, there is an act of the form. In other words the form is such an act as still remains in potency to another act, namely existence. This notion of an act which is itself in potency was very difficult to express in the language of Aristotle. Yet it had to be expressed, since even "those subsisting forms which, because they themselves are forms, do not require a formal cause for both being one and being, do nevertheless require an external acting cause, which gives them to be." In order to receive its to be, a form must needs be in po-

To clarify the difference between "being" and "form", let us quickly draw a few definitions. I say an _un_compounded hypostasis is an individual thing with no potential parts. It is a single, indivisible unit of substrata. In the case of an uncompounded, *material* hypostasis, it is an individual line of force being generated between the actuality and the potency of a single, indivisible unit of substantial substrata. _Un_compounded hypostases receive their essence, and are included in activity, *directly* by means of one of the three kinds of "being" we discussed earlier. In a sense, therefore, "being", in one of its three kinds, is that kind of "form" which *directly* involves *indivisibles* in actuality. To put it another way, "being" is that kind of "form" which actualizes uncompounded hypostases positively, negatively, and neutrally and without the aid of an intermediary. In short, "being" is the "form" of the simple. In this case, however, "form" is not used in its strictest sense.

On the other hand, a *compounded* hypostasis is an individual thing potentially divisible into more than one indivisible part. But, how do these two or more indivisibles manage to be *one* individual thing? The answer is: the one, coherent, mathematical formula which causes each of the several, indivisible units of substantial substrata to develop their six potencies according to the plan of the formula. That formula is a "form" in the strictest sense of that term. In its strictest sense, then, a form is a mediating formula by means of which two or

---

tency to it. "To be," then, is the act of the form, not *qua* form, but *qua* being.
——**ETIENNE GILSON**: *Being And Some Philosophers*; pgs. 174-175. Garden City Press Co-operative. Toronto, Canada, 1949.

more indivisibles become involved in activity as an hypostasis (*i.e.:* as an individual thing).[66]

The simple fact is that, in his **Metaphysics**, even Aristotle himself confesses that *only individual things* can have "being" in the unqualified sense of the word.

> Now these are seen to be more real because there is something definite which underlies them (i.e. the substance or individual), which is implied in such a predicate; for we never use the word 'good' or 'sitting' without implying this. Clearly then it is in virtue of this category that each of the others also *is*. Therefore that which is primarily, i.e. not in a qualified

---

[66] **NOTE OF APRIL 9, 2002:** These days, I no longer use such terminology. These days I use the term "form" to refer to something very much other than "a mediating formula". What "form"—taken in its *strictest* sense—*currently* means to me is thoroughly set forth in **Introspective Cosmology II**; and so, I will not bother to elaborate on it here. These days, I would use "mediating formula" to refer to what I call "*patterns* of divine rotation". These *patterns* are followed by the heavenly *sextets* in their *rhythmical* rotations and by the heavenly *benchmarks* in their *non*-rhythmical rotations. Because each form is associated with one or more sextets and/or benchmarks, the latter serve as the means by which the patterns are imposed upon the forms, and the structure of the imposed pattern determines precisely what kind of material object the form will produce. As Plato might put it: You can have many human beings because many different clusters of matter are—by means of the heavenly sextets and benchmarks associated with the forms in each of those clusters—participating in basically the same pattern of divine rotation devised by God. Oh, how wise were the early "Fathers Of The Church" to prefer Plato to Aristotle!

sense but without qualification, must be substance.

——*Metaphysics*; Book VII; Chap. 1; 1028a: 25-30. ***Great Books***; Vol. 8; pg. 550 bottom left and mid right.

But, the trillions of trillions of atoms racing wildly around in the human body are by no means an *individual* thing, and no matter how closely physics looks at them, physics will not find, either in them or in their juxtaposition, anything which somehow makes them one, *individual* thing. As Aristotle himself puts it in his ***Metaphysics***:

> If we examine we find that the syllable does not consist of the letters + juxtaposition, nor is the house bricks + juxtaposition. And this is right; for the juxtaposition or mixing does not consist of those things of which it is the juxtaposition or mixing. And the same is true in all other cases; e.g. if the threshold is characterized by its position, the position is not constituted by the threshold, but rather the latter is constituted by the former. Nor is man animal + biped, but there must be something besides these, if these are matter,—something which is neither an element in the whole nor a compound, but is the substance; but this people eliminate, and state only the matter. If, then, this is the cause of the thing's being, and if the cause of its being is its substance, they will not be stating the substance itself.

SUBSTRATA

> ——*Metaphysics*; Book VIII; Chap. 3; 1043b: 5-14. **Great Books**. Vol. 8; pgs. 567 bottom right and 568 top left.

By Aristotle's own admission, therefore, there must be something which *bridges the gap* between (1) "being" (which, by Aristotle's own *indirect* admission, is the "form" and the "act" of the simple) and (2) the astronomical numbers of sub-atomic particles racing around in every material object at fantastic speeds in mind boggling patterns. By Aristotle's own admission, that something must be an individual thing, an indestructible formula (see pg. 248, 1st quote), and something which is other than the matter and "which is neither an element in the whole nor a compound, but is the substance." That is why there must be forms, each of which is a simple, indivisible, individual thing which, because its nature is such that it draws many indivisibles together under its common hegemony, *bridges the gap* between the complex and "being" as the "form" and the "act" of the simple.

In conclusion, then, conjunction with a *form* endows the *compounded* material hypostasis with a *compounded* essence, and conjunction with *"being"* endows the _un_compounded hypostasis with an _un_compounded essence. From that, it follows that mankind's struggle to know the world around it must be a three pronged investigation. On the one hand, we must attempt to discover what uncompounded essence is imparted to substrata by its inclusion in substantial activity, and that means we must search for the ultimate, indivisible particle of matter. On the other hand, we must strive to learn the mathematical formula essential to each of the forms.[67] Thirdly, we must endeavor to depict the compounded essence which is imparted to each compounded material hypostasis when matter and a form are united.

For Aristotle, of course, all of that is nonsense, since, for him, there are only *two* principles: matter and form. Consequently, *whatever* matter has—whether being or essence—it can only come to matter by means of the form, and, from that, it follows that all our intellectual efforts should be focused upon the single question: What is the form?

So says Aristotle. But, more importantly, what arguments does he advance to prove that "being" is substance? With what evidences does he convince us that forms alone impart both being and essence? With what compelling words does he show us that each and every individual is equally independent and has the same kind of being? The truth is that there are *no* arguments; there are *no* evidences; and there are *no* compelling words. His assertions are gratuitous ones, and what he freely asserts we can just as freely deny, provided only that we replace his assertions with meaningful alternatives. I will leave it to history to judge whether or not I have done so.

(H)

It is time now to draw this investigation to a close. Before doing so, there is one other topic which must be mentioned.

---

[67] Though I have spent over twenty years trying to deduce these mathematical formulae, I have done no more than to scratch the surface.

SUBSTRATA

In his work **Physics**, Aristotle presents his argument against the idea of indivisibles. I have already written a rather lengthy refutation of his argument and, for that reason, will not dwell on it here. Nevertheless, it does seem appropriate to make a brief remark regarding that argument.

Naturally, Aristotle's argument rests upon several tightly drawn definitions. Those definitions, unfortunately, depend upon, and reflect, Aristotle's gratuitous assumption that matter is a substantial substratum and its motions substantial activities. As a result, all Aristotle really proves is that, if you start off by assuming your conclusion in your opening premises, you can hardly fail to come back to that conclusion in your closing deductions. One has but to distinguish between how Aristotle's terms apply to substantial substrata and how they apply to insubstantial substrata, and his argument collapses dramatically.

In conclusion, then, I allege Aristotle meant to say that the mind of man can conceive of only one kind of substratum. He meant to say that "being"—in so far as it applies to *individual* things—is of only one kind and has only one conceivable definition. He meant to say that, "to be an individual", "to be a substance", "to be a substratum", and "to be a substantial substratum", cannot possibly have but one meaning to the human mind. That is his fundamental fallacy, and I have forever proven him wrong by producing the idea of that which is truly a substratum and which, nevertheless, cannot exist apart from another substratum of a different genus.

Since, however, Aristotle gives no evidences to support his claim that *all* substrata must be *substantial* substrata, neither do I give any evidences to support my contention that *some* substrata are *in*substantial. Therefore, I leave, to the rest of those who covet the title "Philosopher", two questions which *they* must resolve for us all: (1) Is there truly such a thing as an _in_substantial substratum? (2) If there truly is such a thing as an insubstantial substratum, then is matter a *substantial* substratum and its local motions *substantial* activities, or is matter an _in_substantial substratum and its local motions merely the result of the insubstantial activity of substantial substrata?

I am conceited enough to say these two are among the half dozen greatest questions ever to confront the philosophical community, and, consequently, that no man who would call himself a philosopher can possibly ignore them.

It is enough for now. May God grant there is something more in all of this than merely the mindless ramblings of a fool too ignorant to realize how ignorant he is.

Ad Majorem Christi Suaequae Ecclesiae Gloriam! Amen!

His In Omnibus Rebus
Judicium Praevaleat Ecclesiae Catholicae
Quod Est
Judicio Ecclesiae Catholicae
Judicium Cathedrae Petri

ABOUT THE AUTHOR

Born April 13, 1936, in New Orleans, Louisiana, the author graduated from Jesuit high school, in New Orleans, in 1953. A single fruitless semester studying music at Loyola University of the South in New Orleans was followed by almost two years of floundering in a sea of confusion, and the author then joined the U. S. Air Force on Dec. 7, 1955. Honorably discharged in April of 1960, the author underwent another two and a half years of floundering so severe, he came extremely close to a mental breakdown. In desperation, he gave away everything he owned and, for thirteen years, took to the life of a wandering hermit. In search of as much time and energy as possible for inner reflection upon self, God, and the nature and purpose of reality, he criss-crossed the United States on foot four times. At first, he lived off of whatever food and clothing he could beg; but, after learning how to live on a dollar a day or less, he turned to working at various monasteries in the winter time in exchange for the two to three hundred dollars required to feed and to clothe himself during the next spring, summer, and fall of walking. The monasteries also provided access to libraries in which he could read, and extract notes from, the great writings of the Catholic Church. In the course of that thirteen-year odyssey, there was a four year period during which he refused to speak to anyone (except on very rare occasions) and communicated only by means of written notes.

In August of 1975, the author's father lost his mind, and the author's siblings insisted he was the only one in the family with the time and ability to tend to their father in his hour of need. Thus, after thirteen years, the author's *preferred* lifestyle came to an end. Dire poverty then gave way to economic independence, and total seclusion gave way to what little privacy can be enjoyed by bachelors who prefer to avoid partying and to stay home and—as much as possible—to bury themselves in as much reading and writing as the world around them will allow. After his father's death in 1981, the author took care of his mother until her death in 1996.

In this book, the self-educated author of 13 self-published works (No one else would publish them.) seeks to share with others the avenues of thought down which his mind was lead by *thirteen* years of *heroically* intense inner concentration followed by *twenty-two* years of *moderately* intense inner concentration.